BRIGHTER THAN A THOUSAND SUNS

BOOKS BY ROBERT JUNGK

Der Kampf der Schweiz um die Pressefreiheit

(Switzerland's Struggle for Freedom of the Press), 1945

Tomorrow Is Already Here, 1954

Children of the Ashes, 1961

Brighter than a Thousand Suns

A PERSONAL HISTORY OF THE ATOMIC SCIENTISTS

by Robert Jungk

Translated by James Cleugh

A Harvest/HBJ Book
Harcourt Brace Jovanovich, Publishers
San Diego New York London

Hardbound ISBN 0-15-114147-9

Paperbound ISBN 0-15-614150-7

Published originally in Germany under the title *Heller als Tausend Sonnen*

Library of Congress Catalog Card Number: 58-8581
Printed in the United States of America

I J K L M N

For Ruth

Why should we always think of what the scientist does and never what he is?

George N. Shuster, "Good, Evil and Beyond,"
The Annals, *January 1947*

If the radiance of a thousand suns
were to burst into the sky,
that would be
the splendor of the Mighty One—

Bhagavad-Gita

Acknowledgments

As most of the people mentioned in this book are still alive, I have been able to talk to many of them or to obtain information from them in writing. In the course of my researches I have been unable to obtain similar open and uncensored information from Soviet scientists, although I have tried to do so at various international congresses of physicists (in Pisa, Geneva and Rochester). Unfortunately, therefore, this book only deals with the achievements and failures of the West; an unavoidable limitation which I hope will be corrected by future historians. I am responsible for the reproduction or interpretation of all statements cited. In a few cases I have refrained from naming the source of my quotations, at the request of those of my informants who wished to remain anonymous. I am very grateful to the following scientists for devoting so much time to me and showing me such patience.

AUSTRALIA: M. Oliphant.

DENMARK: N. Bohr.

GERMANY: F. Bopp, G. Cario, S. Flügge, W. Gentner, W. Gerlach, O. Hahn, O. Haxel, W. Heisenberg, G. Joos, P. Jordan, H. Korsching, I. Noddack, R. Pohl, M. Schön, F. Strassmann, C. F. von Weizsäcker.

FRANCE: H. von Halban, I. Joliot-Curie, L. Kowarski, Ch. N. Martin.

GREAT BRITAIN: M. Born, O. R. Frisch, K. Furth, K. Lonsdale, R. Peierls, M. Perrin.

JAPAN: N. Fukuda.
AUSTRIA: H. Thirring.
POLAND: L. Infeld.
SWITZERLAND: F. Houtermans, W. Pauli.
UNITED STATES OF AMERICA: H. Agnew, L. Alvarez, H. Bethe, G. Breit, R. Brode, H. Brown, A. H. Compton, C. Daniel, C. Evans, R. Feynman, J. Franck, G. Gamow, S. A. Goudsmit, F. de Hoffman, H. Kalmus, R. Landshoff, R. Lapp, C. Mark, L. Marshall, R. L. Meier, P. Morrison, J. R. Oppenheimer, L. Pauling, V. Paschkis, E. Rabinowitch, A. H. Sturtevant, H. Suess, L. Szilard, E. Teller, G. H. Tenney, V. Weisskopf, N. Wiener, E. Wigner.

I am also indebted to the following, who gave me much encouragement and important information: Fr. M. Bohr, Mrs. R. Brode, Mrs. R. Felt, Mrs. L. Fermi, Fr. M. Hager, Mrs. E. Jette, Mrs. D. McKibben, Mrs. A. Simpson, Mme. A. Vallentin, M. Amrine, J. Bergier, L. Bertin, R. J. C. Butow, H. Chevalier, W. Dames, L. Farago, E. Fuchs, P. Gallois, H. B. Gisevius, L. R. Groves, P. Hein, K. Hirschfeld, A. MacCormack, D. MacDonald, O. Nathan, B. Pregel, R. Reider, A. Sachs, R. Schmidt, A. Schweitzer, K. Selmayr, E. Sommerfeld.

Several other unpublished sources were made available to me:

Files and dossiers concerning the promotion and dismissal of professors in 1933, from the archives of Göttingen University, by courtesy of G. von Selle.

Papers of the Federation of American Scientists in Washington, by courtesy of Miss D. Higinbotham.

Files of the Emergency Committee of the Atomic Scientists in the Harper Memorial Library (Special Collection) at Chicago University, by courtesy of R. Rosenthal.

The testimony of the Japanese atomic scientist Y. Nishina, by courtesy of the Office of Military History, U.S. Army, Washington.

Files of the "Alsos" mission in the possession of S. A. Goudsmit.

Correspondence of Professor A. Sommerfeld, by courtesy of K. Selmayr.

Correspondence between Sommerfeld and Bethe, by courtesy of E. Sommerfeld.

Correspondence about the problem of "self-imposed censorship" (1939), by courtesy of L. Szilard.

Correspondence between Oppenheimer and Chevalier, by courtesy of H. Chevalier.

C. F. von Weizsäcker gave me access to his unpublished comments on *Alsos* by S. A. Goudsmit, as well as "Bemerkungen zur Atombombe" (an uncompleted set of notes made in August, 1945). W. Heisenberg lent me the duplicated text of the parody of *Faust* (Copenhagen, 1932). Pascual Jordan gave me an unpublished manuscript about Heisenberg. Michael Amrine made available to me various notes and articles concerning the "Crusade of the Scientists."

Contents

BRIGHTER THAN A THOUSAND SUNS

One

A Time of Change (*1918-1923*)

It is said that during the last year of the First World War Ernest Rutherford, already famous for his work on atomic research, failed to attend a meeting of the British committee of experts appointed to advise on new systems of defense against enemy submarines. When he was censured for his absence, the vigorous New Zealander retorted without embarrassment:

"Talk softly, please. I have been engaged in experiments which suggest that the atom can be artificially disintegrated. If it is true, it is of far greater importance than a war."

In June 1919, while the attempt was being made in Versailles and other suburbs of Paris to draft peace treaties designed to put an end to the four bloody years of war, Rutherford published in the *Philosophical Magazine* certain studies of his experiments. They showed conclusively that he had succeeded in making an ancient dream of mankind come true. By bombarding the element of nitrogen with tiny alpha particles he had transformed it at various times into oxygen and hydrogen.

The "transmutation of matter" for which the alchemists had searched so long was now a fact. But those precursors of modern science, who took the whole world for their province, considered not only the material but also the moral consequences of such an undertaking. "Deny the powerful and their warriors entry to your workshops," they warned the coming generations of research workers. "For such people misuse the holy mysteries in the service of power."

Rutherford's well-known explanations of the process of transforming the nitrogen atom contain no such warning. It would have transgressed the valued principles of the twentieth century. Philosophical considerations by the modern scientist of the incidental effects of his discoveries would be regarded as improper, even if his studies appeared in the *Philosophical Magazine*. Such has been the rule ever since the scientific academies of the seventeenth century determined that no discussions of political, moral or theological problems should be allowed at their meetings.

As early as 1919, however, the isolation of scientific research had become a working hypothesis only, which actual conditions had already rendered out of date. The war which had just ended had shown only too clearly, by its use of weapons made possible through the practical application of scientific discoveries, the fateful connection between the remote laboratories and the bloodstained realities of the battlefield. Alfred Döblin, the Berlin author later driven by Hitler half round the world, wrote in October 1919: "The decisive assaults upon mankind now proceed from the drawing boards and the laboratory."

Rutherford's workshop, too, had been rudely invaded by war. His "boys," as he called his assistants and students, who looked upon him as a father, had almost been conscripted for military service. Moseley, the most gifted of all his colleagues, had fallen at the Dardanelles in 1915. The source of the radium which he used for all his atomic experiments had been confiscated. It happened to be, by an irony of fate, "enemy alien property."

Before the war, the Vienna Radium Institute had loaned to their highly esteemed British colleague, Rutherford, 250 milligrams of the precious substance. It was a gesture which pre-1914 Austria could easily afford to make; the only productive deposits of uranium ore in Europe lay in the Bohemian district of Joachimsthal, at that time still part of the Imperial and Royal Dual Monarchy. Rutherford never acquiesced in his government's confiscation of the radium loaned him by Austria, nor was he satisfied with the permission of the authorities to use this valuable metal for the time being. He was known to be a scientist of unbending temper and high principles, and insisted on his right to return the personal loan to his colleagues on the Danube at the end of hostilities or else to acquire it from them by paying for it. Rutherford's resolute attitude eventually won out. On April 14, 1921, he was at last able to write to his old colleague, Stefan Meyer, in the inflation-stricken city of Vienna: "I was much disturbed by your statement of the financial side of the Radium Institute of Vienna and have been active in trying to raise some funds to buy at any rate a small quantity of the radium which the Vienna Academy so generously loaned me so long ago and which has been of so much aid in my researches."

Meyer warned that the price of radium in the world market was for the moment "monstrously high." This did not frighten Rutherford. He raised a sum of several hundred pounds which tided the Vienna Radium Institute over the worst years of the currency devaluation.

Even during the war Rutherford had kept in touch, at least by correspondence, through neutral countries with his pupils and friends in Germany and Austria-Hungary, particularly with his old and loyal assistant, Hans Geiger, inventor of the Geiger counter for the measurement of invisible radioactivity, later to prove indispensable. The international family of physicists had kept together to the best of their ability, at all events better than men of letters and intellectuals in other fields, who bombarded each other with spiteful manifestoes. Physicists who had worked together before the war, often for years, either by correspondence or side by side in the laboratory, could never become enemies at a command from above. They helped one another whenever they could. Nernst and Rubens, the German teachers of James Chadwick, helped their pupil set up a small laboratory in the camp at Ruhleben near Berlin, where he had been interned at the beginning of the war. Chadwick, who had been a close associate of Rutherford and was later to win a Nobel prize, carried out a number of interesting experiments with other prisoners in the camp. In May 1918, when the terrible offensives in northern France were daily sacrificing the lives of so many Britons and Germans, he wrote to Rutherford:

"We are now working, or rather about to work, on the formation of carbonyl chloride in light . . . within the last few months I have visited Rubens, Nernst and Warburg. They were extremely willing to help and offered to lend us anything they could. In fact, all kinds of people lent us apparatus."

As soon as regulations at the frontiers were somewhat relaxed, physicists immediately resumed contact to exchange information about the progress made during the years of war. Letters and telegrams were sent to hasten the exchange of information. Telegraph clerks at Copenhagen often found it difficult to pass on correctly the messages, full of mathematical formulae which they did not in the least understand, from the Institute of Professor Niels Bohr to England, France, Holland, Germany, the United States and Japan.

At that time there were three main centers of attraction on the map of atomic research. From Cambridge Rutherford ruled like a sharp-tongued and easily irritated monarch that kingdom of the smallest possible dimensions he had been the first to reveal. Copenhagen decreed through the mouth of the learned Niels Bohr the laws of the bewilderingly new and puzzling territory of the microcosm. Meanwhile Göttingen's triumvirate—Max Born, James Franck and David Hilbert—instantly asked questions about each new discovery made in England and supposed to have been correctly explained in Denmark.

Many fascinating problems presented by the world of atoms could no longer be satisfactorily solved by correspondence. The era of congresses and conferences now began. Bohr only needed to announce that he would lecture in Göttingen for a week on his studies and every physicist tried to make the journey. News of interesting experiments and results achieved came even from lands which before the war either had conducted no physics research at all or only insignificant experiments. India and Japan, the United States and revolutionary Russia tried to exchange scientific information. During these years most zealous efforts were made by the Soviet Union to make contact with Western scientists. The Bolshevist state not only wished its scientists to learn from those "out there." It also took care to have its own publications translated into English, French and German. Even that dictatorial state, in those days, imposed no rules of secrecy or censorship upon the field of research.

A famous physicist remarked at the time that his profession was behaving like a community of ants. Each one hurried excitedly to a breach in the anthill with the tiny fragment of knowledge he had just picked up, only to find that as soon as his back was turned another had taken it away. Planck, Einstein, the Curies, Rutherford and Bohr one after the other administered a series of severe shocks to the edifice of physics, which had seemed at the turn of the century so splendidly easy to survey and so firmly based. Arnold Sommerfeld of the University of Munich, probably the most successful teacher of the postwar generation, said that eager students before they entered the study of physics ought to be warned: "Caution! Dangerous structure! Temporarily closed for complete reconstruction!"

Rutherford unhesitatingly blamed the theoretical, not the experimental, physicists, for the confusion. "They're getting too big for their boots," he growled. "We practical physicists will have to take them down a peg or two."

What had really happened? In the midst of the postwar troubles of the world, its revolutions and inflations, people hardly had the time, the patience or perhaps simply the vitality to grasp the meaning of the most profound of all revolutions, the most significant of all devaluations, the radical change of our image of the world. Planck had shaken the belief, which had been regarded as self-evident for thousands of years, that Nature makes no sudden advances. Einstein had defined as relative the "facts" of space and time until now supposed to be fixed quantities. He had identified matter as "frozen" energy. But now the Curies, Rutherford and Bohr were proving that the indivisible could be divided and that the solid, when one came to scrutinize it precisely, was not stable but in constant motion and change.

Professor Rutherford's alpha particles ought really, at that time, to have upset not only atoms of nitrogen but also the peace of mind of humanity. They ought to have revived the dread of an end of the world, forgotten for many centuries. But in those days all such discoveries seemed to have little to do with the realities of everyday life as men perceived them. The conclusions reached by the physicists through their complicated instruments and their even more complicated calculations about the true character of our world were still, it was generally agreed, solely their own affair. And in fact they themselves did not appear to expect any immediate practical consequences from their revelations. Rutherford had expressly stated his opinion that the world would never be able to exploit the slumbering energy in the atom. It was an error to which he held firm until his death in 1937.

The German physicist and Nobel prize winner Walter Nernst wrote in 1921: "We may say that we are living on an island of guncotton." He was trying to give the latest results of Rutherford's researches greater publicity. Yet he immediately added the consoling comment: "But, thank God, we have not yet found a match that will ignite it."

Why then should man be troubled?

It was true that the physicists themselves were worried. For the time being, actually, they were worrying less about the world than about their own science, in which already hardly any of the old ideas were making sense. But for that very reason much that was new and amazing, much that earlier centuries had never known, was coming to light.

This was the miraculous and exciting era of which one of the youngest, the American Robert Oppenheimer, wrote later:

> Our understanding of atomic physics, of what we call the quantum theory of atomic systems, had its origins at the turn of the century and its great synthesis and resolutions in the nineteen-twenties. It was a heroic time. It was not the doing of any one man. It involved the collaboration of scores of scientists from many different

lands, though from first to last the deeply creative and subtle and critical spirit of Niels Bohr guided, restrained, deepened and finally transmuted the enterprise. It was a period of patient work in the laboratory, of crucial experiments and daring action, of many false starts and many untenable conjectures. It was a time of earnest correspondence and hurried conferences, of debate, criticism and brilliant mathematical improvisation. For those who participated it was a time of creation. There was terror as well as exaltation in their new insight.

Another witness to those years, the great German physicist Pascual Jordan, remembers:

Everyone was filled with such tension that it almost took their breath away. The ice had been broken. . . . It became more and more clear that in this connection we had stumbled upon a quite unexpected and deeply embedded layer of the secrets of Nature. It was evident that wholly new processes of thought, beyond all the previous notions in physics, would be needed to resolve the contradictions—only later recognized as merely apparent—which now came to a head.

Young physicists from all over the world were studying under Sommerfeld in Munich. They even took their problems with them into the cafés. Marble-topped tables were covered with scribbled mathematical formulae. The waiters of the Café Lutz in the Hofgarten, regularly frequented by the Munich physicists, had strict instructions never to wipe the tables without special permission. For if a problem had not been solved by the time the café closed for the night, the further necessary calculations were carried out the following evening. It happened fairly often, moreover, that some unknown person would have the audacity to jot down the solution during the interval. Some young physicist would have been too impatient to wait until the next meeting.

Two

The Beautiful Years (1923-1932)

This tremendous transformation of the scientific view of Nature could only be compared with the change of outlook brought about by Copernicus. It originated, like all really important intellectual revolutions, in places where, to all appearance, deep tranquillity reigned. The most far-reaching revolution of the twentieth century was born in an idyll: a picturesque park in Copenhagen, a quiet side street in Berne, the shore of the island of Heligoland, the meadows and tree-shaded river at Cambridge, the Hofgarten in Munich, the quiet neighborhood of the Panthéon in Paris, the gentle slopes of Zürichberg, and the ancient fortifications of Göttingen, bordered by rustling tall trees.

In the nineteen-twenties Göttingen was the real center of the restless intellectual activities of the physicists. Eminent visitors came from other universities. There were so many of them, especially in the summer months, that the Dutch physicist Ehrenfest observed with his sharp wit:

"We really ought to avoid the rush of our foreign colleagues at the height of the season by visiting, to escape their visits, other establishments of learning."

Between 1920 and 1930 Göttingen was still as dreamy and comfortable as it was in the mid-nineteenth century. To be sure, in this very town in 1908 the first German experimental station for motor travel and aeronautics had been set up. And since the end of the war Göttingen had possessed the first great wind tunnel established in Europe for aerodynamic research. But these laboratories were outside the old town wall and did not change the countenance of the city. The half-timbered houses with their ingeniously carved beams, darkened by the smoke of many, many years, the high Gothic tower of the Jacobikirche, the professors' villas in the Wilhelm Weber Strasse with their twining clematis and glyzinia as in a picture by Spitzweg, the smoky students' taverns, the classically serene Great Hall with its white, gilded pillars, combined to give an impression of something antiquated and consoling that had been preserved during the great war.

The horn of the night watchman continued for many years to sound the day's end, in spite of the radio time signal from Nauen. All kinds of the most modern vehicles might be designed in Göttingen itself, behind the ugly red-brick walls of an institute called in jest the "Lubricating Oil Faculty," but the citizens themselves made far less use of such noisy inventions than the inhabitants of most other German towns.

Most people still went about on foot in Göttingen. The distances to be traversed inside the city were so short that it would have been hardly worth while to go by car or motorcycle. Not until after the First World War did students and professors adopt the bicycle and this was a novelty not popular with everyone. Was it not those leisurely strolls before and after lectures which had so often given rise to the most interesting ideas? Had not chance meetings at a street corner or along the picturesque city wall often accomplished more than formal seminars or committee sessions?

The venerable Georgia Augusta University remained, even after 1918, the spiritual and geographical center of the town. In fact it was more so than ever after the collapse of the old political order. Something of the respect amounting to devotion that had been given under the Empire to higher public officials and Army officers was transferred to the deans and professors of the faculties. The decorations they received, the prizes, degrees and memberships of foreign scientific societies which were bestowed upon them compensated the proud citizens of

Göttingen for the orders and titles conferred in the "good old days."

This respect was also extended, though in smaller measure, to the undergraduates. When the students, especially during the period of intellectual excitement that lasted for the first few years after the war, stood and argued in the streets until late at night, the citizens took it as little amiss as the occasionally noisy homeward journey from the taverns. The landladies of the boardinghouses along the Friedländer Weg, the Nikolausburger Weg or the Düsterer Eichenweg had been accustomed for generations to the students owing them money, though the debts were always paid up in the end somehow or other. Their patience often verged upon self-sacrifice.

One day a young physicist in his first year who had not paid his account at a celebrated bookshop for a long time turned up at the door of the establishment leading a dancing bear, placed at his disposal for the purpose by traveling showmen. The young man, with a perfectly straight face, offered this long-suffering animal to the bookseller "on account." The bookseller, well used to such awkward situations, did not turn a hair. He acknowledged with a laugh that it was really he—not the bear—who was being led round by the nose.

The retired professors were treated like princes. They were assured of the veneration of all. Though they no longer gave lectures, they continued to take a lively part in the intellectual life of the town. Most of them still belonged to and often presided over its scientific bodies. The best seats were reserved for them when lecturers came to visit. When the old gentlemen took their leisurely walks along the streets which in some cases already bore their own names, they were respectfully greeted and even asked for advice, now by a young colleague who might be seated at his open window preparing a forthcoming lecture, now by a tutor, younger still, just arrived by invitation from some other university, and seated on a bench absorbed in scribbling ideas in a notebook. The steady progress of science and the acquisition of knowledge went on and no outside disturbance seemed able to trouble it.

Never before—and perhaps never again—would university men have so much cause to consider themselves the true leaders of society as here in Göttingen during the "beautiful years." In the Ratskeller is an old students' motto: *Extra Gottingam non est vita.* To many members of the university who studied, taught or spent the evening of their lives in the city it must have seemed that the statement was daily confirmed anew.

Eminent philologists, philosophers, theologians, biologists and professors of law had contributed to the establishment of the worldwide fame of the Georgia Augusta. But the University of Göttingen owed its celebrity above all to its mathematicians. Carl Friedrich Gauss had taught there until about the middle of the nineteenth century. He had made Göttingen the center for the most abstract of all sciences. From 1886 on a man sat in this renowned chair who had consolidated and perhaps even enhanced its reputation by his standing not only as a thinker but as a bold, tireless and inspired organizer. He was Felix Klein.

For nearly thirty years, from 1886 to 1913, Klein worked in Göttingen. He was a tall, upstanding man with bright, penetrating eyes. He was described by the daughter of the mathematician Carl Runge as "having something kingly about him." A journey to America in 1893 had impelled Klein to try to abolish the distinction, at that time strictly maintained in Europe, between pure science and its various applications. He labored incessantly to insure that mathematics should "keep in closer touch with practical life."

It was Klein, therefore, who really initiated the foundation or further extension of many astronomical, physical, technical and mechanical institutes in Göttingen. There gradually grew up round them, in consequence, a whole private industry for the production of scientific measuring apparatus and optical precision instruments. The old-fashioned town became a cradle of the most modern technology.

It is characteristic of the liberal tradition that still prevailed in those days that Klein did not hesitate to invite to Göttingen mathematicians like Hilbert and Minkowski who were utterly opposed to him intellectually. These men uncompromisingly repudiated every kind of specialization and all attempts, however tentative, to make any practical application of mathematics. Hilbert's lofty spirit, exclusively concentrated upon ultimate essences, could feel only contempt for "technicians." On a certain occasion he had to replace Klein, who was ill, as the faculty member in charge of students of the Mathematical School at the annual engineers' congresses at Hanover which Klein had arranged. Hilbert was warned that he must give a conciliatory lecture and speak against the notion that science and technology were irreconcilable. Bearing this admonition in mind, he declared in the somewhat harsh East Prussian dialect he affected:

"One hears a lot of talk about the hostility between scientists and engineers. I don't believe in any such thing. In fact I am quite certain it is untrue. There can't possibly be anything in it because neither side has anything to do with the other."

Dozens of such anecdotes about Hilbert, whose frankness amounted to positive gruffness, circulated in Göttingen, but no one resented his ironic malice and well-aimed gibes. They expressed the same inflexible honesty with which he approached his own field of mathematics. This trait permitted him to proceed repeatedly to the most unexpected conclusions without paying the slightest attention to intellectual conventions. With justice his lectures attracted students from all over the world. When he stood by the enormous slide rule that overlooked his desk and raised the still-unsolved problems of mathematics, all who listened felt that they were taking a direct part in the revelation of new knowledge. His hearers did not leave the lecture room with dead facts long proved, but with living questions.

There was only one problem from whose solution Hilbert deliberately abstained, though he might have been able to win a small fortune—a hundred thousand gold marks—by elucidating it. This was the sum which a learned citizen of Darmstadt had bequeathed to anyone who could find the answer to a certain mathematical problem, "Fermat's Last Theorem," which had remained unsolved ever since the seventeenth century. So long as no correct solution came to light the trustees of the bequest were entitled to devote the interest on the fund to any object they chose. It was used every year to enable prominent mathematicians and physicists to hold a series of lectures in Göttingen. Henri Poincaré, H. A. Lorentz, Arnold Sommerfeld, Planck, Debye, Nernst, Niels Bohr and the Russian Smoluchowski were among those thus invited to Göttingen. All made invaluable suggestions. "It's lucky that I am probably the only person who can crack that nut," Hilbert used to say every time he glanced through, in his capacity as chairman of the prize committee, the attempts at solutions annually submitted by laymen and professional mathematicians, pronouncing them all, as usual, inadequate. "But I shall take very good care not to kill the goose that lays us such splendid golden eggs."

Every Thursday, at three o'clock sharp in the afternoon, the four masters of the Mathematical Institute, Klein, Runge, Minkowski and Hilbert, gathered in the porch overlooking the garden of Hilbert's home. A big blackboard stood there, half in the open. Hilbert had often been scribbling on it right up to the last moment, as the chalk on the sleeves of his jacket would show. The discussion of a new series of formulae often began there and then. It would continue while the participants climbed through the woods and over the open fields in all weathers up to the "Kehr" hotel on the heights. There, over coffee, the illustrious quartet would argue about all the small and large questions of their private lives, their beloved university and the wide world. Again and again, as the talk went on, often reaching the most rarefied atmosphere of the limits of human understanding, loud laughter would intervene, giving comfort and relaxation to minds that had reached the seemingly unconquerable frontiers.

One of the many important new institutions which Felix Klein's inventive gift for organization had bestowed upon Göttingen was the mathematical reading room in the Auditorium building. It contained not only the leading mathematical and physical periodicals of the world, as well as a reference library of manuals, but also summaries and occasionally even the entire typescripts of current lectures. Teachers and students who possessed the keys to both rooms could work in perfect peace between lectures and also—which often turned out to be more important—discuss the subjects of their reading in the anteroom, where the strict rule of silence did not have to be observed. Debates between physicists and mathematicians had never ceased since modern developments in natural science had required the help of mathematics to express their contradictory perceptions. Hilbert had remarked in his usual irritable style: "This will never do! Physics is obviously much too difficult for physicists!"

He was not content with uttering that negative opinion. With characteristic zest he took up the study of what he used to call the "intellectually poverty-stricken" science of physics and attempted to give mathematical assistance.

Probably due to his influence one of the most gifted theoretical physicists of the "new school" was invited to Göttingen in 1921. Max Born, at that time just thirty-eight years old, was no stranger to the Georgia Augusta. The son of a well-known Breslau biologist, he had graduated at Göttingen in 1907 as one of the most brilliant pupils of the Mathematical Institute, receiving a prize for his work. His studies and travels took him to Cambridge, Breslau, Berlin and Frankfurt. With his arrival at the Second Physics Institute in the Bunsenstrasse—a red brick building with an unspeakably homely exterior like a Prussian cavalry barracks—the brief but incessantly productive golden age of Göttingen atomic physics began.

A small bureaucratic error, one of those tricks of fate which can accomplish so much, helped Born soon after his arrival. Although a chair for experimental physics already existed in Göttingen, its occupant, Professor Pohl, spent practically all his time teaching and had far too little leisure for the research to which Born was looking forward. But the new head of the institute discovered on examining its papers that provision had been made in the budget for a second chair which had never been filled. This was merely due to a clerical error, he was told. Born refused to give in and insisted on the letter of the law. He was able, therefore, to call to Göttingen James Franck, already well known for his experimental discoveries, including the one which later gained him his Nobel prize.

Hilbert, Born and Franck, a trio of men of high talent, tireless industry and a fervent passion for the new view of Nature, worked together in Göttingen after 1921. Each differed fundamentally from the others. Born was probably the most interested in the outside world, the most accessible and the most versatile. His talents were so various that he might well have become a first-rate pianist or author. His wealthy father had given him the following advice before he began his studies: "Be sure you try out all the courses before you decide which one to follow." So he entered his name simultaneously, during his first semesters at the university at Breslau, for lectures on law, literature, psychology, political economy and astronomy. He felt himself most attracted to the last because, so he said, he preferred to all the other buildings that in which lectures on the world of stars were given!

Franck, like Born, came of a Jewish family which had long been settled in Germany. He could never forget his Hamburg origin. In spite of the cordiality and warmth which made him very popular with his students, he kept other people at a certain distance. He remained always a Hamburg aristocrat. "A distinguished fellow," people called him in those days. Later those who worked with him called him a saint. This was not only because of Franck's great kindness of heart but also because of his almost religious devotion to physics. He would tell his pupils that only one who was entirely absorbed by physics and actually dreamed about it could hope for enlightenment. He spoke of his own inspirations in the language of a medieval mystic. "The only way I can tell that a new idea is really important is the feeling of terror that seizes me."

In almost every age a certain sphere of human reflection and creative activity exercises a peculiar fascination for gifted minds. In one period restless seekers after novelty take a special interest in architecture. In others they apply themselves to painting or music, theology or philosophy. Suddenly, no one knows how it happens, the most alert spirits perceive where new ground has recently been broken and press forward eagerly to become, not only its heirs, but its founders and masters.

Atomic physics exercised this magnetic power in the years after the First World War. It was taken up by the philosophically talented, by men with artistic gifts, by politically minded young men who were repelled by the confusion of day-to-day politics, and by adventurous spirits who could find no more to conquer in a world whose most distant continents had been explored. Disclosures were still possible in the study of the most invisible and microscopic of all phenomena. Here one might come upon traces of new laws, and might experience the peculiar delight, mingled with fear, of having thought something which no one had yet thought, of having seen something which no one had yet seen.

Since there was so much that was new and uncertain in the domain of atomic research, teachers and pupils drew closer together than in other disciplines. Experience and knowledge were worth little. Old and young became comrades on this journey into the interior of matter. Both alike took pride in their common conquest of fragments of knowledge. Both showed equal modesty and bewilderment before the impenetrable.

James Franck, who already held the Nobel prize for physics, could turn from the blackboard on which he had lost his way in a difficult calculation and inquire of one of his students, "Perhaps *you* can see the next step?" The professors made no secret of their mistakes and doubts. They kept their pupils posted on their private correspondence where unsolved problems were discussed with their foreign colleagues, and encouraged their youthful collaborators to seek for explanations which their elders had been denied.

A highlight of every week of the term was the "Seminar on Matter" conducted in Room 204 of the Institute by Born, Franck and Hilbert, *gratis et privatissime.* It became almost a tradition for Hilbert to open the proceedings with a pretense of innocence: "Well, now, gentlemen, I'd just like you to tell me, what exactly *is* an atom?"

Each time a different student tried to enlighten the professor. The problem was tackled afresh every time, and every time they searched for a different solution. But whenever any of the young geniuses sought refuge on the esoteric heights of complicated mathematical explanations, Hilbert would interrupt him in broadest East Prussian: "I just can't understand you, young man. Now tell me over again, will you?" Everyone was forced to express himself as clearly as possible, and to build solid bridges across the gaps of knowledge instead of trying to jump them with overhasty strides of thought.

These debates were concerned more and more with the most basic problems of epistemology. Had the discoveries of atomic physics abolished the duality between the human observer and the world observed? Was there no longer any real distinction between subject and object? Could two mutually exclusive propositions on the same topic both be regarded as correct from a loftier standpoint? Would one be justified in abandoning the view that the foundation of physics is the close connection of cause and effect? But in that case could there ever be any such thing as laws of Nature? Could any reliable scientific forecasts ever be made?

Questions, questions and still more questions. They could be discussed without end and everyone had something to say about them. In the winter semester of 1926 a slender, rather delicate-looking American student distinguished himself, even among such highly talented people as these. He was often able to improvise on the spur of the moment entire dissertations, so that hardly anyone else had a chance to speak. At first the new boy was listened to with fascination. But after a time his excessive garrulity and eloquence began to cause irritation and possibly also envy among a number of his companions. They submitted a written petition to one of the professors suggesting that a check might be put on the *wunderkind*. In a little less than twenty years he was to become world famous: J. Robert Oppenheimer, who was introduced to the public for the first time by the newspapers of August 1945 as the "father of the atom bomb."

Oppenheimer was one of the many young Americans who came to the Old World in those years to study physics. They sometimes called themselves "Knights of Columbus in reverse," for they traveled in the opposite direction to that taken by Columbus and those who accompanied him. They, too, were in search of a "new continent." They returned from it to their own country, where "old-fashioned physics" was still being taught, bringing back incredible information and fabulous discoveries which, like the gold procured by the Spanish seafarers of the sixteenth century, were to prove of great but troublesome profit to their native land.

Almost all these young Americans came to Europe richly endowed with traveling scholarships. They were joined by some older hands, teachers who were in the habit of spending their sabbatical year—twelve months of private study on full pay awarded by tradition every seven years—as learners, exchanging ideas with their European colleagues.

These scientific tourists from the other side of the Atlantic brought foreign currency into the university towns of Europe, impoverished by the war. Further capital in the shape of dollars often followed them, for the American university men were frequently successful in pleading the cause of their temporary European alma mater and in obtaining funds from philanthropic organizations.

The impoverished German scientific institutes in particular, which perpetually complained of lack of money, benefited very much from this American assistance. What would Privy

Councillor Sommerfeld at Munich have done without the occasional improvement in his scanty resources provided by the Rockefeller Foundation? Whenever Wickliff Rose, the distributor of funds endowed by the oil magnate, traveled in Europe, the universities received him like a despot. On the size of his check depended the number of research programs which could be continued during the ensuing year and the number of young research workers who could be given scholarships.

The American mathematicians and physicists were particularly fond of Göttingen. Before the First World War Charles Michelson had worked there for a term as visiting professor, and Millikan and Langmuir, the grand old men of American physics and chemistry, had studied at Göttingen.

In the nineteen-twenties there were often a dozen or more Americans enrolled in the faculty of natural science at the Georgia Augusta. They brought with them to Göttingen a little of the unburdened atmosphere of the American campus. Their annual Thanksgiving dinners, the most memorable presided over by K. T. Compton in 1926, were universally popular. The Americans showed their German colleagues how to eat turkey and sweet corn and learned in turn to drink beer, to sing and to hike. Nearly all the Americans who became well known later on for the development of atomic energy had been at Göttingen at various times between 1924 and 1932. They included Condon, who complained in lively fashion of the lack of comfort in the Göttingen lodgings; the lightning-brained Norbert Wiener; Brode, always deep in thought; the modest Richtmyer; the cheerful Pauling—one of Sommerfeld's pupils, who often came over from Munich; and the amazing "Oppie," who managed to pursue in Göttingen not only his physical studies but also his philosophical, philological and literary hobbies. He was particularly deep in Dante's *Inferno* and in long evening walks along the railway tracks leading from the freight station would discuss with colleagues the reason why Dante had located the eternal quest in hell instead of in paradise.

One evening Paul Dirac, who was usually so silent, took Oppenheimer aside and gently reproached him. "I hear," he said, "that you write poetry as well as working at physics. How on earth can you do two such things at once? In science one tries to tell people, in such a way as to be understood by everyone, something that no one ever knew before. But in the case of poetry it's the exact opposite!"

Oppenheimer and Dirac both lived in a fine granite villa at the start of the Geismarer Landstrasse, facing the Astronomical Observatory where Carl Friedrich Gauss had once worked. The villa belonged to a medical man, Dr. Cario, whose son Günther was preparing for a brilliant future as a physicist while acting as one of Franck's assistants. It was a usual practice for Göttingen families of good social position to take in students as "paying guests." They brought the outside world into the provincial parlor and received in return a measure of domestic security which they smiled at at first but soon came to value and look back on with longing. Between those who leased the rooms and those who rented them often grew long-standing friendship and occasionally marriage. A surprising number of the wives of professors on the five continents come from little Göttingen.

From these families the foreign students often learned German very quickly. They frequently even wrote articles in German for scientific periodicals during the period of their studies. In conversation, however, they made amusing mistakes. The young English astrophysicist Robertson wanted one day to check the exact weight of a letter he was going to send abroad. He burst into a shop and breathlessly asked the girl behind the counter: *"Haben Sie eine Wiege? Ich mochte etwas wagen."* ("Have you a cradle? I want to do something risky.") The girl blushed and stared at him and he hastily corrected himself. "I beg your pardon, I meant to say *haben Sie eine Waage? Ich mochte etwas wiegen."* ("Have you a pair of scales? I want to weigh something.")

American students could never get used to the bureaucratic formalities common in German universities. Even Oppenheimer stumbled over them. In the spring of 1927 he applied for permission to take the examination for a doctor's degree. To everyone's astonishment his request was flatly refused by the Prussian Ministry of Education, to which the university of Göttingen was subordinated. An inquiry by the dean of the Natural Science Faculty elicited the following reply from Ministerial Councillor von Rottenburg in Berlin:

"Herr Oppenheimer made a wholly inadequate application. Obviously the Ministry had to refuse it."

Oppie had apparently forgotten to comply with the regulations which required a detailed account of his career along with his application for admission to the Georgia Augusta. He had never formally matriculated, therefore, and consequently had never been a member of the university at all!

The professors of the future father of the atom bomb had to write imploring letters to the Rectorial Board and the Ministry. Max Born said Oppenheimer's work for his doctor's degree had been so outstanding that Born wanted to publish it in one of the series of Göttingen dissertations. The plea that the American undergraduate could not wait another term at Göttingen to take his examination in regular fashion was expressed to the authorities in a petition for the grant of belated matriculation. "Economic circumstances render it impossible for Herr Oppenheimer to remain in Göttingen after the end of the summer term," it declared.

Was this argument justified by the facts? Oppenheimer was the son of a New York businessman who had left Germany for the United States at the age of seventeen and made a fortune there. Consequently it was not so much money that Oppie lacked, in all probability, as patience. He was bound to regard a further term at Göttingen as a waste of time. In these years, however, little fibs of this kind had not yet become the subjects of committees of investigation. The petition went through without further objection.

Robert Oppenheimer took his oral examination on the afternoon of May 11, 1927. He passed in all subjects—except physicochemistry—with the marks "excellent" or "very good." His written work for the doctorate was pronounced by Max Born to be evidence of a high grade of scientific achievement, far above the level of the average dissertation. Born's only criticism stated that "the one fault to be found with the work is that it is difficult to read. But this formal shortcoming counts for so little in comparison with the content that I propose the paper be marked 'with distinction.' "

In the Göttingen of these twenty years it was possible to get along without a scholarship or a fat monthly check. The Russian mathematician Schnirelmann brought nothing with him but his toothbrush and a copy of his latest work on prime numbers, but the Göttingen mathematician Landau had already given a lecture on the "Schnirelmann proposition," and the young scholar, who had arrived in rags, soon got some decent clothes and board and lodging at the well-known Pension Wunderlich, frequented by physicists. His anonymous patrons also sent him every month a small money order to cover out-of-pocket expenses. He was often seen slowly making his way down the main street of Göttingen, his shoelaces undone and his gaze, as usual, fixed absent-mindedly upon distant combinations of figures.

The eminent stage director Kurt Hirschfeld, who was studying in Göttingen at the time, tells us how odd he thought these young mathematicians and physicists were. He once saw a member of Born's "kindergarten," who was walking along in a dream, stumble and fall flat on his face. Hirschfeld rushed up and tried to help him to his feet. But the fallen student, still on the ground, vigorously repulsed his efforts. "Leave me alone, will you? I'm busy!" Apparently a new brilliant solution had just occurred to him. Fritz Houtermans, today a professor of physics at a Swiss university, reports that he was once awakened in the middle of the night by someone banging on the window of his ground-floor rooms in Nikolausburg Strasse. One of his fellow students was begging urgently for admission. He had just had a splendid idea, he said, which would dispose of some of the unsolved contradictions in the new theories. Far from driving the intruder from his door, the disturbed sleeper opened it, as soon as he had got into his dressing gown and slippers. The two worked until dawn on calculations with the newly established equations.

In those exciting years it was not unusual for such sudden "brain waves" by very young men to make a great stir in international professional circles or even in certain cases to bring their authors fame almost overnight.

For example, Werner Heisenberg, the son of a professor of ecclesiastical history, had spent his last year at school in the throes of the Munich "Councils" Revolution* and served with an anti-communist unit of schoolboys. To bring food to his starving family in the blockaded city he had twice, at the risk of his life, slipped through the lines of "White" and "Red" troops. While on sentry duty on the roof of a theological seminary he had read Plato and been aroused by the atomic theories of the ancient Greeks. But the opinion stated in Plato's *Timaeus* that atoms were ordinary substantive bodies satisfied him as little as a drawing in his physics textbook which depicted them with hooks and eyes. This critical attitude of refusal to be impressed by any authority did not desert Heisenberg even when his instructor, Sommerfeld, took him to Göttingen in 1921 to attend the Bohr Festival Season. Far from merely listening with reverence to the great man from Copenhagen, the boy, who was then only just nineteen years old, crossed swords with him during long walks to the Rohns and up to the Hainberg.

Because of these conversations, which delighted Heisenberg, he decided to study physics. His name would soon be read as a collaborator in one of Sommerfeld's publications. At twenty-three he was working as assistant to Born, at twenty-four he was lecturing on theoretical physics in Copenhagen and at twenty-six he became a regular professor at Leipzig. When he was barely thirty-two years old Heisenberg received the Nobel prize for theoretical studies of fundamental importance, published actually some six or eight years previously, in other words at an age when most students of medicine and law have just concluded their training.

* *Translator's note: i.e.*, the Communist "Soldiers' Councils," or "Soviets," established after the Armistice of 1918.

One of his closest friends states in his recollections of Heisenberg at Göttingen: "He looked even greener in those days than he really was, for, being a member of the Youth Movement, the moral idealism of which greatly attracted him, he often wore, even after reaching man's estate, an open shirt and walking shorts. He always considered himself constitutionally lucky and this was quite true. Brilliant intellectual achievements, such as his recognition of the 'uncertainty principle' or the basic ideas of the 'matrix calculus,' which he afterwards developed with the help of Born and his fellow student Pascual Jordan, a few months younger than himself, seemed simply to have fallen into his lap.

"Those who came to know Heisenberg later, after the political upheaval had troubled him with grief and doubt, cannot know how radiant he once was. He had brought his revolutionary quantum mechanics with him in 1925 from Heligoland, where he used to climb on the red cliffs while he read Goethe's *West-Östlicher Divan* and worked in the intervals on his own ideas in a kind of intellectual intoxication. I doubt if he had any sleep worth mentioning during that blissful Whitsun vacation."

The lean and lanky Dirac, son of a Swiss father and an English mother, attained a high reputation in the world of physicists when he was even younger than Heisenberg. Even the initiated could not always follow his mental processes. The "mystic of the atom" was not in the least worried by that. When he was not at Cambridge he could often be seen working in one of the classrooms of the Second Physics Institute at Göttingen. As if in a dream, he would be holding a mental colloquy with the rows of symbols he had chalked on the blackboard. Even in the presence of a second person Dirac hardly ever accompanied the stages of his mathematical proofs with words. Speech of course would never have been able to express what he had to say. The other physicists used to say that Dirac was so silent that he uttered an entire sentence only once every light year.*

* In one of the usual sketches produced at the end of term in Eurcpean universities the author attributed the following quatrain to Dirac:

This little group of young men between twenty and thirty years old, inspired by or endowed with genius, included Enrico Fermi, an Italian who had attracted hardly any attention among his gifted colleagues in Göttingen. Another was "Pat" Blackett, a former British naval officer, who photographed and interpreted the miraculous world of atomic events. That merry and freakish soul from Soviet Russia, George Gamow, had more ideas than anyone else but left it to others to distinguish the true from the false. Then there was Wolfgang Pauli, from Vienna, who once danced for joy in the middle of the Amalienstrasse at Munich because something new had just occurred to him. They all knew, of course, that they were engaged on work of far-reaching importance. But they never dreamed that their somewhat esoteric studies would so soon, so profoundly and so violently affect the fate of mankind and their own lives.

The young Austrian Houtermans could never have suspected at that time that certain ideas which he advanced one hot summer day in 1927 during a walking tour near Göttingen with his fellow student Atkinson would lead a quarter of a century later to the explosion of the first hydrogen bomb, the "absolute" weapon that might be instrumental in bringing about an end of the world.

To pass the time the two undergraduates had raised partly as a joke the old, unsolved problem of the true source of the inexhaustible energy supplied by the sun that was beating down above their heads. There could be no question in this case of any ordinary process of combustion, otherwise the substance of the sun must long since have been consumed in the fierce heat generated for so many millions of years. But ever since Einstein's formula of the interchangeability of matter and energy the suspicion had been growing that in all probability a process of atomic transmutation was at work in that tremendous laboratory in the sky.

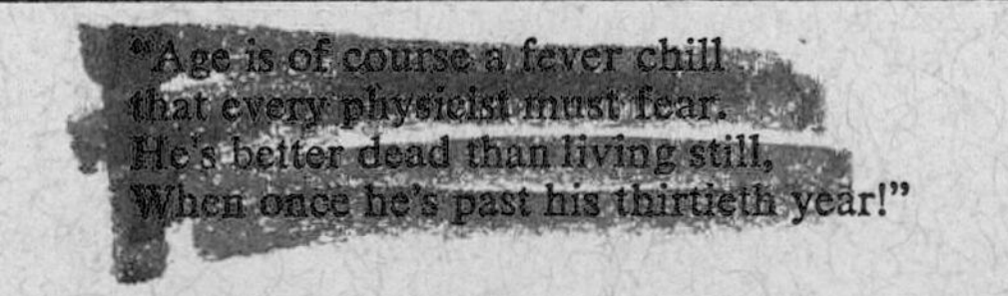
"Age is of course a fever chill
that every physicist must fear.
He's better dead than living still,
When once he's past his thirtieth year!"

Atkinson had participated in Rutherford's transformations of the atom at Cambridge. He suggested to his companion that what had been accomplished in the Cavendish Laboratory must also be possible "up there."

"Naturally!" Houtermans retorted. "Let's just work the thing out, shall we? How could it happen in the case of the sun?"

Such was the origin of the labors of Atkinson and Houtermans on their theory of thermonuclear reactions in the sun, which later achieved such fame. The theory for the first time put forward the conjecture that solar energy might be attributed, not to the demolition, but to the fusion, of lightweight atoms. The development of this idea led straight to the H-bombs that threaten humanity today.

At that time, of course, neither of the two young students of the atom dreamed of such sinister consequences.

Houtermans reports: "That evening, after we had finished our essay, I went for a walk with a pretty girl. As soon as it grew dark the stars came out, one after another, in all their splendor. 'Don't they sparkle beautifully?' cried my companion. But I simply stuck out my chest and said proudly: 'I've known since yesterday why it is they sparkle.' She didn't seem in the least moved by this statement. Perhaps, she didn't believe it. At that moment, probably, she felt no interest whatever in the matter."

Three

Collision with Politics (1932-1933)

A textile manufacturer named Levin had leased the second floor of a villa to James Franck, in the Merckelstrasse at Göttingen. One evening a foreign physicist, as so often before, was a guest in the drawing room. But on this occasion he was listened to with more than usual attention, for Professor Abraham Joffe came from Soviet Russia and had wonderful reports of the practical support given to scientists by the state. There they had no financial worries like those the Second Physics Institute had to contend with. The rooms at the Institute were hardly heated at all during the cold winter of 1929. At late seasons of the year, in order to save electricity, it was strictly forbidden to work before ten in the morning or after four in the afternoon. The man from Leningrad declared that at his own institute there were three hundred students and numerous well-paid assistants. They could all count on safe employment and steady promotion, for their aspiring country needed competent scientists.

"All that sounds good," said Joffe. Then he lowered his voice suddenly and added in a barely audible tone: "I often feel as if I were living on top of a volcano. One never knows when, nor for what reason, it is actually going to erupt."

The internal struggle for power between Stalin and the other factions of the Communist Party in the USSR had become so acute that it had begun to affect even the scientists. Certain Soviet physicists, who had previously been able to travel abroad without much difficulty, suddenly and inexplicably ceased to appear. The few who did visit the rest of Europe began noticeably to keep their distance. The Soviet physicist Landau, who was suspected of Trotskyism, begged his colleagues at the Berlin Technical Academy not to discuss politics with him on any account. Only a few years earlier the exact opposite had been the case. Landau himself had defended the new order of society with fiery zeal. He had come to Berlin with holes in his shoes and couldn't understand at all why he was given not only one but two pairs of new ones. "Who can possibly need two pairs of shoes?" he said, criticizing such "capitalist customs."

It was rumored in Göttingen, Cambridge and Copenhagen that Gamow, the "innocent" among the physicists, who had always been ready to amuse his listeners with conjuring tricks and childish games, was just then playing a decidedly less hilarious version of hide-and-seek with the secret police. When he found that he was no longer to be allowed to visit the West, he had at first made a vain attempt to escape across the Afghan border. He was caught, but convinced the frontier guards who picked him up that he was only mountain climbing. Gamow, who had recently married, soon made a second attempt to escape. He planned to cross the Black Sea, with his wife, in a small sailing vessel, from the Russian to the Turkish shore. Unluckily he ran into so violent a storm that he was glad to be rescued from imminent shipwreck by the motorboats of the frontier police patrol, the very people he had intended to evade.

And yet Gamow belonged, actually, to the post-Revolutionary generation, neither for nor against Communism, but simply ready to take advantage of the opportunities afforded by the state for further education and the making of a career. Now, however, the new lords of the "machine" in Moscow had indicated that they were no longer content with the neutral attitude of scientific specialists. They demanded active ideological support.

Modern physics in particular, in the form it had assumed in the West during the twenties, appeared ideologically suspect to the cultural commissars of Soviet Russia. The assertion made, above all by Bohr and Heisenberg, that in the act of observing submicroscopic processes it was no longer possible to draw any clear distinction between the subject (that which observed) and the object (that which was observed) clashed with the doctrine of materialism. For this view allowed the individual far too much influence over natural phenomena. Such a concession amounted in the eyes of the official philosophers of the Soviet Union to "dangerous idealism" which could only end in "ecclesiastical obscurantism."

Jaroslav Frenkel, in a "lecture for the toilers" in Moscow, gave an account of the theory that light, according to the conditions under which it was observed, could at various times be described as consisting either of small particles or, nearly, of waves. He then added, by way of a joke: "Of course, according to the type of thinking hitherto prevalent among us, these two alternatives absolutely exclude each other. So, Comrades, you can believe in the particles on Mondays, Wednesdays and Fridays and in the waves on Tuesdays, Thursdays and Saturdays." A woman from the audience seized upon this remark. She administered a stern rebuke to the lecturer for indulging in "bourgeois propaganda." The authorities got wind of the incident and the physicist was persecuted as a "reactionary," though all he had done was to report on the latest advances in his professional field. Even the article under Frenkel's name in the Soviet Encyclopedia did not fail to censure Soviet Russia's possibly greatest contemporary teacher of physics: "The philosophical ideas of Ja. L. Frenkel are not notable for their lucidity and consistency so far as his attitude to materialism is concerned. Many of the statements in his books suffer directly or indirectly from idealist distortion and have been rightly subjected to strong criticism by the community of Soviet scientists."

The tranquillity even of Göttingen was disturbed, when the economic crisis began in 1930, by louder and louder echoes of the grating notes of politics. The leading newspaper of the city, the *Göttinger Tageblatt,* a middle-class journal, had followed an extremely conservative policy for some years. It started to praise Adolf Hitler as a redeemer at a time when the rest of the nationalist press in Germany still qualified their judgment of the "leader" with certain reservations. Students of the Second Physics and the Mathematical Institutes were discreetly combining into a National Socialist group. For the moment they confined themselves to spreading anti-Semitic propaganda among their followers, from which their own Jewish professors were of course exempt, since their characters, after all, were known from daily experience to be above all reproach. There was also a small but active Communist cell among the Göttingen students smuggling Communist leaflets and pamphlets into the library of the Physics School. The matter was investigated. Although the culprits were not discovered, the atmosphere of the Institute, formerly so friendly, grew tense and suspicious.

Some years before, nationalist students had hissed Einstein off the platform in Berlin from which he was lecturing on his relativity theory. The incident had disgusted people in Göttingen. "Here," they said, "we have only once had a demonstration by students. They marched out to the house of a famous physicist who had just arrived and gave him an enthusiastic welcome by chanting, in the twilight, Planck's quantum formula." Now, however, demonstrations were beginning to occur rather often, even in this idyllic university city, against "objectionable" instructors like the outstanding mathematician Hermann Weyl, who was a close friend of Einstein.

The brown-shirted students made a particular onslaught against Jewish or half-Jewish undergraduates who had come from Poland or Hungary to study in Germany. These people were already victims of the "cold" anti-Semitism of their native lands, which had denied them admission to universities under the *numerus clausus* law restricting the number of Jewish university students to a small quota. Now they were sacrificed for the second time to racial prejudice. Talented young natural scientists such as Eugen Wigner, Leo Szilard, John von Neumann and Edward Teller were at that time making notable contributions in Göttingen, Hamburg and Berlin to discussions on the subject of atomic physics. Only a few years later they became the most active champions of the construction of the atom bomb. The alarm which they then felt at the possibility that Hitler might be the first to possess so terrifying a weapon can only be understood when one realizes what abuse and persecution they had to endure from National Socialist students in 1932 and 1933. They were never really able to get over the shock of the inroad of political fanaticism upon the peace of academic life, a shock which was destined to make history.

Long before Hitler's seizure of power a small group of German physicists, styling themselves "national researchers," had formed around the Nobel prize winners Lenard and Stark. This group boldly declared Einstein's theory of relativity to be "Jewish world-bluff." They attempted to dismiss, under the summary heading of "Jewish physics," all studies based on the data of Einstein and Bohr. Even at that time they characterized as "Jewish-minded" the Aryans who founded their published work on relativity and quantum mechanics. Johannes Stark was especially bitter against Sommerfeld. This arrogant inventor of a nebulous science of "German physics" had been offended by the pertinent criticism directed at his work by the Munich professor who had dared to call him in jest "Giovanni Fortissimo."* The nickname, planted first by Einstein, had stuck ever since. Stark also held his eminent Munich colleague responsible for his dismissal from the University of Würzburg. In reality Stark had been fired because he had used his Nobel prize money for the purchase, contrary to the statutes of the Stockholm foundation, of a china factory, and thereafter had taken more interest in this than in his scientific duties. The learned world of the Weimar Republic did not take very seriously these excursions of a few of their number into the foggy regions of demagogic racialism. For the time being professional achievement counted more than anything else. The adherents of "German physics" who had become agitators no longer attracted much attention, and their "stupid clamor" was regarded as of no importance.

* *Translator's note: Stark* in German means "strong."

The growing agitation of the cranks, the unrecognized and the unsuccessful who gathered about the National Socialist physicists, was in fact a symptom of the profound political and social unrest in Germany. Unemployment statistics rose week by week. Every day the newspapers reported meeting-hall riots between "brown shirts" and members of other political parties. Political assassination became a commonplace event. But the Göttingen atomic physicists, like most other natural scientists throughout the world, at first behaved as though it were possible to ignore these outrageous events. With an obstinacy amounting to monomania they applied themselves even more intensively than before to their work.

The reasons for this ostrich-like policy were clearly analyzed by Franck fifteen years later. It was in 1947 that he told the Emergency Committee of Atomic Scientists:

> "It is a custom in science—and perhaps a principle—to select from the infinite reservoir of unsolved problems only those simple ones the solution of which seems possible in terms of available knowledge and skills. We are trained to subject our results to the most severe criticism. Adherence to these two principles results in our knowing very little, but on the other hand being very certain that we know this little.
>
> "We scientists seem to be unable to apply these principles to the immensely complex problems of the political world and its social order. In general we are cautious and therefore tolerant and disinclined to accept total solutions. Our very objectivity prevents us from taking a strong stand in political differences, in which the right is never on one side. So we took the easiest way out and hid in our ivory tower. We felt that neither the good nor the evil applications were our responsibility."

The fame of the University of Göttingen had slowly grown over centuries of patient, brilliant achievement and studies crowned with success. It had spread throughout the world. A few months, in fact only weeks, in the spring of 1933, were enough to destroy its reputation. At the Georgia Augusta, and at most other German seats of learning, there were noisy demonstrations by a minority of students who pretended to represent a majority. There were fiery speeches by political demagogues to proclaim the coming of the "new order." There were brutal expulsions of respected scholars whose opinions or heredity were flung in their faces like crimes. In Göttingen's retreat this seemed even more senseless and outrageous than in other university cities. For there one knew one's fellow citizens too well to believe in the incessant flood of accusations from the new masters of the state. It was known for certain that the men who were being given notice to quit their posts were irreplaceable. Pupils from all Europe, from the United States and even from Asia had come to sit at their feet. If they left, Göttingen would sink to the level of a provincial establishment.

Almost a hundred years before, seven professors had been obliged to leave the University of Göttingen because they had protested against the violation of the Constitution by the King of Hanover. Now also, seven professors, the first victims of another violation of the Constitution, were compelled to leave. For barely a month after Hitler's seizure of power telegraphic orders were received from Berlin for the immediate retirement of seven members of the Natural Science Faculty. Most of them, Max Born, for instance, who happened to be abroad at the time, made no serious effort to contest this arbitrary decision. Only one member, the mathematician Courant, tried to resist the decree by submitting petitions. But neither his reference to the fact that he considered himself entitled to be treated as a "patriotic German" after being shot in the stomach and badly gassed as a front-line soldier at Verdun in the First World War, nor the petition of protest on his behalf signed by twenty-two German professors—including Heisenberg, Hilbert, Prandtl and Sommerfeld, as well as the Nobel prize winners von Laue and Planck—helped matters in the slightest.

Franck was at first exempted, no doubt because, as a Nobel prize winner, he was particularly well known abroad. But he had sufficient pride to dispense voluntarily with any such special privilege. On April 17, 1933 he sent in his resignation. Two days later he informed the public, through the few newspapers which had not yet been rendered completely subservient to the Party line, that he felt obliged to withdraw in sympathy with his expelled colleagues. "We Germans of Jewish origin are now treated like aliens and enemies of our country," he complained, stressing the fact that he had no desire to occupy any special position.

The great physicist's honorable attitude was, however, taken amiss by certain professors of the Georgia Augusta. Instead of defending academic freedom and intellectual dignity forty-two lecturers and professors forwarded an infamous document to the headquarters of the local Göttingen group of the National Socialist Party. In this communication they condemned Franck's action as "playing into the hands of foreign atrocity propaganda." Only a single one of Göttingen's natural scientists had the courage to protest openly against the dismissal of the Jewish savants. This was the physiologist Krayer. He did not allow himself to be intimidated either by his own dismissal, which was then ordered by the new Prussian Minister of Education, Stuckart, or by the threat of being debarred from employment for the rest of his life.

The great majority of the Göttingen professors deplored the invasion of their quiet retreat by demagogues and hatred, but to protect their professorial chairs, they dared no protest. When second- and third-raters, whose only merit was to have joined the Party in time, began to reorganize everything and issue decrees, they met, instead of real resistance, only a lightly mocking attitude, which harmed nobody. The new National Socialist *Dozentenführer** played first fiddle in those days, as a purifying agent and representative of the new order, but was soon unmasked as the plagiarist of other people's work. Though he was recognized by many, even at that early stage, as a thief of intellectul property and a braggart, no one found the civic courage to demand his recall. The intention was to co-operate in order to rescue what could still be salvaged. Through this policy the remaining professors became more or less disguised

* *Translator's note:* The *Dozentenführer*—a post created by the Nazis—was the political supervisor of the lecturers attached to each university.

supporters of a regime destined to bring untold misfortune on the country it administered and on the world.

Some weeks after these melancholy events the colleagues, pupils and friends of Franck, gathered once more in the tea-room of the Second Physics Institute to bid him farewell. On the eve of their chief's departure they wished to assure him of their gratitude and esteem. His assistant Cario handed him, with a short speech, a portfolio of pictures of Göttingen to take with him on his travels as a memento of the beautiful years. Franck was visibly moved as he acknowledged the gift.

Next day he vacated his rooms in the Merckelstrasse villa. He drove to the station unaccompanied, for he had asked to be allowed to leave alone. Nevertheless, a number of his friends came to the station. Ahlborn, the porter, described Franck's departure in the following words: "Just imagine," he said. "After the Herr Professor had got into the train, the thing didn't move. Engine didn't want to go. Had more sense than our new leaders!"

Those left behind in Göttingen—there were still some eminent men of learning among them, even then—could never rival, in their subsequent labors under the Third Reich, the great achievements of the twenties. The clearest account of the state of the university was given by the mathematician Hilbert, by that time well advanced in years. About a year after the great purge of Göttingen he was seated at a banquet in the place of honor next to the new Minister of Education, Rust. Rust was unwary enough to ask: "Is it really true, Professor, that your Institute suffered so much from the departure of the Jews and their friends?" Hilbert snapped back, as coolly as ever: "Suffered? No, it didn't suffer, Herr Minister. It just doesn't exist any more!"

There remained one island of peace and mutual tolerance in the flood of noisy political fanaticism. At the *Universitets Institut for Teoretisk Fysik,* No. 15 Blegdamsvej, Copenhagen, physicists of all nations, races and ideologies gathered, as in the years before Hitler's seizure of power and Stalin's "new line," around their master Niels Bohr. The more shamelessly the pretensions of half-truths and lies spread from one nation to another in public life, the more strenuously the people in Bohr's circle labored to discover the obscure image of the whole truth, perpetually elusive, forever withdrawing to deeper and deeper strata. The new dictators did not tolerate anything outside the items in their programs and inflicted savage punishment upon the mildest critics of their plans. The "spirit of Copenhagen," on the other hand, invited criticism. It insisted upon the observation of everything from several points of view and postulated the eventual synthesis on another level which appeared to be contradictory.

The supposedly unworldly Bohr took more rapid and effective action than any other member of the family of physicists to help those of his colleagues who were living under the dictatorships. Many of those engaged in atomic research who were still resident in Germany found sudden and urgent invitations from Bohr in their letter boxes, though they had made no previous application for any such thing. "Come and stay with us for a while," such messages would run, "and think things over quietly until you decide where you would like to go."

Those who, in the autumn of 1933, arrived at Copenhagen by the afternoon train from Germany, and found members of Bohr's Institute waiting to welcome them as cordially as their own relatives, felt after a few hours that a miracle had occurred in their return from the world of Party decrees and mute anguish to the formerly familiar atmosphere of reciprocal esteem and friendship.

Bohr lacked, as his pupil Weizsäcker put it, two qualities which usually distinguish most heads of a school. He was neither a pedagogue nor a tyrant. He showed no signs of offended pride when his ideas were sternly or even rudely criticized. The free-and-easy relations between master and pupils in an establishment headed by Bohr are vividly represented in a parody of *Faust* which was staged at the beginning of the thirties on the occasion of the annual September School attended by Bohr's foreign students. The Lord in this play is obviously intended to be Bohr himself. The part of Mephisto was allotted to his pupil and tireless critic Wolfgang Pauli. One conversation between these two, given in slightly exaggerated terms for the sake of the parody, was reported as follows:

BOHR (the Lord):
Hast thou naught else to say?
Comest thou ever with complaint?
Is physics never to thy mind?
PAULI (Mephisto):
Nay, 'tis all folly! Rotten, as ever, to the core!
E'en in my days of dule it grieves me sore
and I must ever plague these physicists the more.
B. (*In the mixture of German and English he always used when excited*): Oh, it is dreadful! In this situation we must remember the essential failure of classical concepts . . . *muss ich sagen* . . . just a little remark . . . what do you propose to do with mass?
P.: What's that got to do with it? Mass? We shall abolish it!
B.: Well, that's very, very interesting . . . but . . . but——
P.: No, shut up! Stop talking rubbish!
B.: But . . . but——
P.: I forbid you to speak!
B.: But Pauli! Pauli! We're really much more in agreement than you think! Of course, I quite agree! Only . . . Certainly, we can abolish mass. But we must uphold load. . . .
P.: What for? Why? No, no, that's wishful thinking! Why not abolish load, too?
B.: I must ask . . . I understand perfectly, of course . . . but . . . but——
P.: Silence!
B.: But Pauli, you must really give me a chance to finish what I have to say! If both mass and load are abolished, what have we got left?
P.: Oh, that's quite simple! What we've got left will be the neutron! (*Pause. Both pace to and fro.*)
B.: It's not to criticize, it's but to learn
I take my leave now, later to return.
(*Exit.*)

P. (*Soliloquizes*):

I like to see the old chap now and then
and take good care we don't fall out. It's jolly
decent of so grand a Lord, I must say, when
he comes himself for a nice chat with Pauli!
(*Exit.*)

Bohr did not take himself very seriously, with the result that slight lapses of respect for his personality did not seem to vex him. Perhaps for that very reason all who worked with him held him in such great honor and felt such warm affection for him as has seldom been the lot of any teacher. They smiled at his absent-mindedness and failure to remember things, but in the very act of smiling they admired a spirit capable of overlooking for the time being all external matters in order to concentrate on the really essential. In 1932 the government had placed at his disposal, to show its appreciation of the most learned man in Denmark, the castle of Carlsberg. When Bohr left the castle on his bicycle for the Institute, he hardly ever took any notice of red traffic lights. If he traveled by trolley he was often so deep in thought that he not only passed his stop and went on to the terminal, but occasionally forgot again to get out at the same stop on the return journey. And yet Bohr's activities were by no means confined to study. He went sailing with his pupils, carved windmills for them, helped them to solve crossword puzzles and played ping-pong with them. But his favorite game was always soccer. As a young man he had been a useful member of a good team. It was true that some frivolous people declared that he would sometimes, instead of kicking the ball towards the opposing goal, pick it up in the middle of the game in order to see precisely what that sphere of leather might really contain.

Niels Bohr was a poor speaker. Nearly all his lectures began with similar sentences, in which he would for the hundredth time give his reasons for departing from classical theory. His pupils called this exordium "Bohr's celebration of Mass." He often spoke in too low a tone, mixed up German, Danish and English expressions, and put his hand in front of his mouth during the most important passages. In mathematical knowledge he was greatly inferior to the majority of his audience. Yet what he had to say went deeper and meant more, if one took the trouble to develop its significance, than most of what could be heard from physics professors at other universities whose lucidity and impressive powers of oratory far exceeded Bohr's.

Bohr's true greatness was most evident to his pupils in private conversation. When a new piece of work was submitted to him, his first comment was usually: "Magnificent!" But it was only newcomers who exulted prematurely over this exclamation. Those who had known Bohr longer were well aware that, for instance, the words: "Very, very interesting—" uttered with a slight, deprecating smile after the lecture of a visiting professor, really denoted a most contemptuous verdict. By asking certain questions, occasionally speaking at length or else remaining silent for some minutes, the great thinker would gradually persuade the young physicist who had come to him for advice to realize for himself that perhaps his work had not yet been quite thoroughly thought out. Such an interview might last for several hours and be prolonged until late at night. Frau Bohr would sometimes come quietly and unostentatiously into the room. The students admired her domestic virtues even more, possibly, than they did her classic beauty. Without a word, at most with a smile, she would hand the disputants plates of excellent sandwiches, together with a few boxes of matches for the lighting of the master's pipe, which was always going out.

In the end the pupil would not only begin to discover the faults in his work but would also start remorselessly tearing it to pieces himself. At this point, however, Bohr would check him, warning the young man against indiscriminate rejection, since even errors contained something which might later turn out to be useful.

"When, after a few years, one left his Institute, one knew something about physics which one didn't know before and couldn't have learned in any other way," Weizsäcker once said in talking of Bohr. It was no wonder that so many eminent natural scientists emerged from his classes. He was one of those rare teachers who know how to apply caution and, if necessary, force, to the task of liberating the slumbering genius in each man's mind. Like Socrates, whose expression of thought through dialogue he considered exemplary, Bohr was a midwife of ideas.

Among those who studied with him in Copenhagen during the year of crisis brought about by Hitler's seizure of power were two unusually different friends. Carl Friedrich von Weizsäcker was the highly gifted son of a prominent German diplomat. Edward Teller was a Hungarian who had been forced to leave Germany on account of Hitler's racial decrees. The friendship between the German aristocrat and the expelled non-Aryan was rare because Weizsäcker, like so many young German idealists, mistakenly believed at that time that possibly Hitler and his movement, despite certain features of it which Weizsäcker himself repudiated, might be the preliminary symptoms of something really admirable, the beginning of a social and religious revival in revolt against the spirit of commerce and arid intellectualism. He made no secret of his hopes in this respect and at first would not allow the skeptical Teller to convince him of the opposite view. He repeatedly argued that it might be one's duty to find some good in the regime of which he had come to know the dark side only too well in Copenhagen, the refuge of so many victims of National Socialism. Five years later these political discussions, which really only took place on the margin, so to speak, of conversations on physics and general philosophy, were to have an important influence on the course of global history. For in 1939 a small group of physicists, including Teller, who had emigrated to the United States, learned that Weizsäcker was supposed to be in charge of the German "uranium project." Teller, among others, thereupon urged the American military authorities to undertake the construction of an atom bomb as a preventive measure. He assumed that his former fellow student, in his admiration for the political successes achieved by Hitler's power politics, would support him in spite of the horror with which he had at times regarded the Führer. In reality, as it happened, Weizsäcker had by that time entirely got over his illusions about National Socialism, though only his most intimate friends in Germany knew it.

Teller, the son of a respected Budapest lawyer, had known ever since his tenth year that he would never be able to make a career in his native land, whose laws excluded nearly all Jews from academic education. Accordingly, at eighteen he left Hungary for Karlsruhe, where he studied chemistry. As a result of the interest he soon began to take in the quantum theory he planned to continue his studies under Sommerfeld in Munich. But he saw little more of the Bavarian capital than the four walls of a hospital. Teller, who was a keen mountaineer, one Sunday morning in 1928 after his arrival in the city, was on his way to catch an excursion train bound for the Alps. Finding that he was rather late he alighted from the moving trolley in front of the main station so hurriedly that he was knocked down and his right foot amputated. "I don't have much luck in this town," he thought, so he went to Leipzig. Heisenberg had just been appointed a professor there and was starting to collect a band of talented young pupils. It was at Leipzig that Teller first met the dreamy, highly imaginative Weizsäcker, who was four years younger than he. Weizsäcker had really intended to study general philosophy, but in Copenhagen, where his father had been sent by the Foreign Office on diplomatic service, he met Heisenberg, working then with Bohr. "One can't get anywhere in philosophy now without knowing something about modern physics," Heisenberg told him at a party given by his father. "But you'll have to start on physics pretty soon if you don't want to be too late."

Teller himself had always been gifted with a highly imaginative character. Only a few people guessed that he wrote poems in secret. The friendship between him and Weizsäcker was founded less upon their common interest in science than on their love of poetry, literature and philosophical speculation. After Teller had obtained his doctor's degree at Leipzig he went to Göttingen to work under Born, in collaboration with whom he wrote a study of optics. After Hitler came to power Teller escaped to Copenhagen by way of London. He had married a girl he had known when he was a boy, but he kept the marriage secret—the Rockefeller scholarship he held was awarded only to bachelors. He lived as the unmarried men who worked at Bohr's Institute did, residing in one of the private boardinghouses nearby. Two of these places were especially popular with the physicists, those run by Fröken Have and Fröken Thalbitzer. It was a matter for debate which of these two ladies was the more extraordinary character. The first had picked up so much mathematics over the years from her learned guests that she was fond of expounding to them her own theories of heaven and earth, while the other declared all such hairsplitting to be morbid, smoked a pipe, often wore an old soldier's cap, and told the young men they ought to throw their "silly books" into the sea. "I love to hear the waves roar," she used to exclaim in her deep voice, when she returned from one of her frequent excursions to the beach. "That's where you learn to know Nature! Not from dry, printed papers!"

Teller and Weizsäcker both lodged with this energetic lover of nature. Weizsäcker was in the habit of going to Teller's room about midnight for friendly disputes that lasted till two o'clock in the morning. They enjoyed close argument so much that Teller invented a discussion game. From time to time one of them had to try to convince the other of the truth of an utterly paradoxical statement. In the ensuing debate one of the partners was only allowed to indicate his position by putting adroit questions in the style of the Platonic dialogues. One of the propositions which Weizsäcker tried to prove in those days was, "To stand at attention is a Dionysian experience," while Teller advanced the following: "Malicious pleasure is the purest of pleasures."

Another game which had a serious background consisted of the questionnaires they submitted to each other. Teller, later the "inventor" of the most dreadful weapon in the world, gave certain answers at that time which are particularly interesting in the light of what he was to become. To the question, with what sort of things would he least like to be concerned, he replied, not quite twenty years before the great hydrogen-bomb experiments, "Machines." The next question was: "What do you most like doing?" Teller answered: "Making clear to others what they find obscure and obscuring what they find clear." Teller, who later composed for the American government so many confidential reports of great influence on the future course of history, stated that what he most hated doing was "writing for other people." In reply to a further question he compared himself, among characters in fairy tales, with Lucky John, whose gold nugget melted away to nothing, and among historical personages with Louis II, a king of Hungary whom he much admired for staking everything on a single card at the battle of Mohacs and thus losing both his kingdom and his life.

Teller often used to translate into German verses by the Hungarian poet Ady, a great favorite of his. One of these translations, which were never published, seems full of the presentiment of an existence which Teller and other research workers in atomic physics were fated to lead.

> The Lord takes all whom he strikes and loves,
> he bears them far from Earth . . .
> their hearts aflame, their brains made ice,
> Earth sends her laughter up to them
> and, compassionate, the sun strews diamond dust
> upon their lonely way.

Copenhagen could be only a temporary refuge for the many atomic-research scientists who had escaped from Central Europe. The indefatigable Bohr, with Lord Rutherford's powerful assistance—the latter had been created a baron in 1931—repeatedly managed to procure means of subsistence for his colleagues who had been robbed of their appointments, incomes and savings. But this life of dependence pleased no one. It could not go on forever, and posts were sought for the expelled physicists throughout the world. This proved to be less easy than it had appeared, for the number of vacant chairs in

European universities and the space available for work in the laboratories were limited. Hardly any country yet understood that the admission of refugees who brought with them no material possessions but invaluable knowledge would be profitable rather than a burden.

Only the United States, with its hundreds of universities and institutes, could provide enough appointments for the intellectual refugees. In the first two years after Hitler's accession to power the United States was still suffering from the consequences of the great economic crisis that had begun in 1929. But when Albert Einstein, in the autumn of 1933, accepted the offer of employment at the newly established Institute for Advanced Study at Princeton and transferred his residence from Berlin to that small American university city, the French physicist Paul Langevin, half in jest and half in earnest, made a true prophecy. "It's as important an event," he said, "as would be the transfer of the Vatican from Rome to the New World. The Pope of Physics has moved and the United States will now become the center of the natural sciences."

It was surprising that the Russians, formerly always so anxious to engage foreign specialists, made so little effort to admit the expelled scientists. The fact was that Stalin's promotion of Soviet patriotism and the growing fear of sabotage and spies had already begun to make the curtain between the Soviet Union and the rest of the world more and more impenetrable. A provisional corner, however, was lifted to admit foreign "comrades." Accordingly, a few well-known atomic-research physicists did emigrate, as "fellow travelers" or members of the Party, to the Soviet Union. A terrible fate eventually overtook them there. Until 1937 they were allowed to work in physics institutes and even to publish a periodical in German. But during that year they fell victims, in common with other foreign Communists, to the great purge. They were thrown into prison and tortured. Attempts were made to extract false confessions from them. They were sent to Siberia.

The West at first suspected nothing of all this. Bohr actually sent to the Soviet Union, as late as 1938, two of his pupils, the Viennese physicists Weisskopf and Placzek, to investigate the prospects of employment for the new refugees coming from Austria, just occupied by Hitler. The two emissaries were regarded by Stalin's secret police as espionage scouts and could only report to Copenhagen, in the end, the sad tidings of the fate of the physicists Houtermans and Alexander Weissberg.

Petitions signed by Einstein, Bohr and Joliot-Curie—the latter only became a Communist some years afterwards—were then forwarded to the Kremlin. They probably saved the lives of the two Western scientists. Houtermans and Weissberg were at first offered Soviet citizenship and restoration to all the offices they had formerly occupied, but they refused this offer because of their earlier appalling sufferings. In 1940, the Soviet secret police escorted the scientists, who had originally fled from Hitler, to the frontier of the "General-gouvernements" —Poland as then occupied by German troops. There the two men were delivered up by the Ogpu to their "colleagues" of the Gestapo, only to face new trials and new imprisonment.

Four

An Unexpected Discovery (1932-1939)

At the beginning of the thirties, during the same period in which politics had so brutally invaded the quiet world of the laboratories, nuclear science also knocked at the door of politics. In 1932 James Chadwick discovered the neutron, the key to atomic fission.*

But the knock was a very gentle one. Hardly anyone heard it. Fritz Houtermans stated in his 1932 inaugural address to the Technical Academy in Berlin that this tiny particle, which had then just been discovered at Cambridge, might well be able to free powerful forces dormant in matter. His statement attracted little serious attention. Three years later Frédéric Joliot-Curie went to Stockholm to receive with his wife, Irène, the Nobel prize for their discovery of artificial radioactivity.

He said there: "We are justified in reflecting that scientists who can construct and demolish elements at will may also be capable of causing nuclear transformations of an explosive character. . . . If the propagation of such transformations in matter can be brought about, in all probability vast quantities of useful energy will be released."

Even Joliot's prophetic words aroused no more than a transitory interest. Only one professional investigator drew, al-

* A neutron is a microphysical elementary particle which can penetrate the atomic nucleus in spite of the latter's protection by electric barriers of millions of volts, because the neutron is electrically neutral and is therefore not repelled.

most immediately, political conclusions from the scientific revelation provided by the discovery of the neutron. The Hungarian physicist Leo Szilard, born two years before the turn of the century, had already suffered, as quite a young man, the effects of political convulsions. He had been barely a year at the Technical Academy in Budapest before he was conscripted. The war was going badly for the Triple Alliance, but the Imperial and Royal officers still drilled their recruits as sternly as in the years of the great reviews before the Emperor. They gave Szilard so passionate an aversion to everything military that nearly thirty years later he replied to an American reporter who asked him what his hobby was: "Beating brass hats!"

The weeks of the Red Terror under Bela Kun, followed by the months of the White Terror under Horthy, drove Szilard, who had at first tried to continue his studies in Budapest after demobilization, to Berlin. There he began by joining the Technical Academy at Charlottenburg, and in the following year matriculated at the university. Einstein, Nernst, von Laue and Planck were then teaching and working in the German capital. Under their influence Szilard, who had originally intended to become a civil engineer like his father, turned to theoretical physics. The vivacious, imaginative young scientist soon began to make a name for himself in his chosen field, first as von Laue's assistant, then as an unsalaried lecturer taking part in the work of the Kaiser Wilhelm Institute.

When Hitler achieved power Szilard first went to Vienna, but his special gift for deducing future events from present facts made him realize that Austria would sooner or later be overrun by National Socialism. He therefore stayed only six weeks in Vienna before leaving for England.

In the autumn of 1933, at the annual meeting of the British Association, Lord Rutherford made a speech in which he observed that people who were prophesying the release of atomic energy on a large scale were "talking moonshine." "This set me thinking," Szilard recollects, "and it occurred to me in October 1933 that a chain reaction might be set up if an element could be found that would emit two neutrons when it swallowed one neutron. At first I suspected beryllium, then some other elements, including uranium. But for some reason or other the crucial experiment was never carried out."

Szilard was not, in fact, experimenting any further with neutrons at that time. He was wondering, as a realistic student of future events, how politicians, industrialists and soldiers would be likely to react if success were really achieved one day in the generation of atomic energy. It was true that hitherto no one had been able to penetrate to the hard core of the atom and harness its slumbering energy to practical purposes. But so many investigators and their teams were already working at such problems that it seemed the solution might not be far off. As soon as this discovery, at present "floating in the air," could be brought down to earth once and for all, the current indifference of the governments in power would be certain to turn to alert attention. Ought not scientists to realize that possibility here and now?

Such considerations caused Szilard, as early as 1935, to approach a number of atomic-research workers and ask them whether it would not be advisable, in view of the possibly momentous and perhaps even dangerous consequences of their present studies, to refrain, at least for the time being, from publishing any future results of their investigations. His suggestion was for the most part repudiated. After all, there seemed to be no chance of the fortress of the atom ever being conquered. And yet Szilard was already talking of what should be done with the prize! Because of this "premature" proposal he acquired the reputation of always thinking about third and fourth steps before the first and second had been taken.

However, other scientists, too, were troubled by similar venturesome speculations. Paul Langevin, who did so much in those years for refugees from the Third Reich, was sufficiently stirred by these ideas to attempt to console, in a somewhat unconventional manner, a student of history who had made his escape from Germany. "You're taking it all much too seriously," said he. "Hitler? It won't be long before he breaks his neck like all other tyrants. I'm much more worried about something else. It is something which, if it gets into the wrong hands, can do the world a good deal more damage than that fool who will sooner or later go to the dogs. It is something which —unlike him—we shall never be able to get rid of: I mean the neutron."

The young historian had so far heard only quite casual reports about this neutron supposed to be so dangerous. That was the business of a different faculty, he told himself, and had nothing to do with him. At that time he understood as little as most of his fellow men that great scientific discoveries may have a more lasting effect upon history than mighty dictators.

The underestimation of politics by men of science was surpassed at that time, a quarter of a century ago, only by the underestimation of science among politicians and the public. If the number of times that the name of Hitler cropped up in those days were to be compared statistically with the number of times the word "neutron" was mentioned, the ratio of a million to one would probably put the figure too low. So little are we ourselves, even in the "age of information," able to judge which contemporary events may prove in the end to be of real importance and an omen for the future.

Only since the end of 1945, when the whole world was made aware of the discovery and technical development of atomic power, has it been possible to see the fission of the atom as a turning point in global history. What extraordinary coincidence that in the same twelve months the neutron was discovered (February 1932), Roosevelt was elected (November 1932) and Hitler took over the German government (January 1933).

Seven fateful years were now to pass before physicists recognized the full significance of the neutron, seven years in which atoms had already been split with neutrons in Paris, Cambridge, Rome, Zürich and Berlin without anyone suspecting the fact. The scientists themselves were not aware of it. From 1932 until the end of 1938 they simply refused to believe what their instruments told them, and therefore the statesmen in their turn fortunately did not learn the nature of the extraordinarily powerful weapon that lay within their reach. It is interesting to speculate what the consequences would have been if the chain reaction in uranium had been correctly interpreted, in Rome, in 1934, when it probably took place for the first time. Would Mussolini and Hitler then have been the first to develop an atom bomb? Would the atomic-armaments race have begun before the Second World War? Would that war, we may wonder, have eventually been fought out with atomic weapons on both sides?

The physicist Emilio Segré had taken part in those successful but wrongly interpreted experiments in the Italian capital. He later made an attempt to answer such questions, which he, like many other specialists in atomic research, had often asked himself. Two decades afterwards, at the funeral of his teacher Enrico Fermi, he said: "God, for His own inscrutable reasons, made everyone blind at that time to the phenomenon of nuclear fission."

It was no accident that the discovery of the neutron occurred in Rutherford's laboratory at Cambridge rather than elsewhere. At the Zürich Physics Congress in 1931 the announcement by the Germans Bothe and Becker that they had observed a very strong, inexplicable radiation* on bombarding beryllium with alpha particles caused a very great sensation. Investigators in all countries immediately attempted to repeat the experiment to discover the nature of the radiation mentioned. Joliot-Curie and his wife solved the puzzle to some extent. No later than a month after they had published their first results, Chadwick, who had been working at this problem almost uninterruptedly, encouraged by Rutherford, announced that neutrons were involved. Their existence had been predicted by Rutherford no less than seventeen years previously.

Chadwick owed his rapid success in large part to the superiority of his measuring instruments, in particular a new amplifier which had just been invented. In 1932 no physics-research

* The "gamma rays."

establishment in the world possessed such excellent instruments as the Cavendish Laboratory at Cambridge.

Atomic physicists can never perceive and examine the objects of their studies with the naked eye. Only scientific equipment can make those objects perceptible and measurable by the use of apparatus. These implements, without the aid of which no human action is practicable in the realm of the smallest possible dimensions, were still extremely primitive towards the end of the First World War. Investigators were accustomed to concocting them, with the help of their laboratory assistants, out of wire, wax and glass apparatus which they blew themselves. The further they tried to penetrate into the unknown the heavier and more complex such equipment became. In 1919 C. D. Ellis, the English physicist, saw for the first time the experimental apparatus with which Rutherford had just performed his first transmutations of atoms. He wrote later: "The whole apparatus was contained in a small brass box and the scintillations were viewed with a microscope. I can remember being surprised, in fact mildly shocked, that the apparatus was not more impressive." Less than fifteen years afterwards Ellis himself, who had meanwhile joined Rutherford's "stable" at the Cavendish Laboratory, hardly ever used for his experiments anything but huge generators and new, highly sensitive measuring instruments. Workshops for practical atomic research began more and more to resemble factory assembly halls and the specialists themselves engineers.

These new galleries of instruments were naturally expensive. Up to the end of the First World War the Cavendish Laboratory had never spent more than £550 a year on new apparatus. Gradually this figure rose until by the thirties it was already several times that amount. This brought about by degrees a change in the relation of atomic scientists to society at large, though it was only much later that they themselves became aware of it. Funds had fomerly been provided annually by wealthy individuals for the growing expenses of the laboratories: by the Canadian tobacco trader McGill (who, by the way, thought smoking a frightful habit and forbade it in the laboratories he financed), the Belgian manufacturer of chemicals Ernest Solvay, and the large-scale German industrialist Carl Still, known as the "good angel of the Göttingen physicists." These gifts were no longer considered sufficient, and even the millions of the Rockefellers, Mellons and Austins now fell short. State intervention was found more and more necessary. Some governments, the British, for instance, were ready enough to help. Others proved more reluctant. In cases where public assistance had originally been insignificant the atomic specialists often applied with eventual success for larger subsidies. In those years it never occurred to them that their new patron, the state, might one day say: "He who pays the piper calls the tune."

Since the Cavendish Laboratory was at that time technically so much better equipped than any other experimental station in the world, atomic physicists expected the discovery of the neutron to be followed by further important news about the character of and the effects produced by this nuclear constituent.

Their expectation was even more justified because Rutherford had managed to collect an altogether exceptional team of collaborators.

One was the somewhat melancholy Aston, builder in 1919 of the prototype of the mass spectrograph, with which he first measured the mass of separate isotopes. Another was the invariably good-humored Japanese Shimizu, whose new "cloud chamber" automatically photographed the paths of atoms. P. M. S. Blackett was outstanding among the rest. A tall, rather stiff type of naval officer, he had only come to the Cavendish Laboratory to take a holiday course, but became so fascinated by the new studies that he stayed on to prove himself the most successful cartographer of the newly conquered territories through his interpretation of 440,000 atom tracks on cloud-chamber photographs.

Nor should one forget the temperamental Australian Marcus Oliphant, nor that unrivaled expert in the handling of the new electrical instruments, John Cockroft, nor Norman Feather, famous, in particular, for his almost superhuman patience. These men worked as a kind of subsection directed by the Russian physicist Pyotr L. Kapitza. In 1921 Kapitza had left his native land, then in the throes of civil war and famine, to join Rutherford at Cambridge. The Kapitza Club, consisting of some twenty young men, met once a week, outside the Laboratory, for a scientific bull session. On these occasions, Hans Bethe reports, Kapitza used to ask every two minutes, "Why is that so?"

All these young students of the atom were possessed by intense zeal for their work. Rutherford called them simply his "boys"; he often treated them like a strict schoolmaster, but the truth was that he loved them like a father. Having no son himself he lavished all the vigilance, help and affection he had to give on these aspiring young men. Whenever he suspected that one of his "lads" was on the way to bringing some new discovery to light, he fussed over him from morning till night and even telephoned the Laboratory late at night to give him advice and warm encouragement as he sat watching his instruments.

For a long time Kapitza was undoubtedly Rutherford's favorite. He admired the Russian's stubborn obstinacy, coupled with great mental agility, the speed at which "Peter" could work, and his delight, bordering on the fanatical, in being busy. But above all Rutherford felt in Kapitza a kindred spirit, although he was almost twenty-five years his senior. It used to be said of Rutherford: "He's a savage, a noble savage, perhaps, but still a savage." Or again: "Relations with Rutherford are not ordinary. One couldn't be friendly with a force of Nature." Such observations also applied to Kapitza. He had the same capacity as his master for enthusiastic enjoyment of life, the same unbridled energy and imaginative power—to which he added, in the bargain, a streak of Russian eccentricity.

Whether he was motoring at top speed along the quiet English country roads, leaping naked into a neighboring stream, as he did at weekends to the disgust of his puritanical hosts—frightening the swans with his imitation of rooks cawing—or spending several nights without sleep while he experimented, like a god hurling thunderbolts with a high-frequency generator, increasing the load till he set one of the cables on fire, this son of a tsarist general invariably lived beyond the borderline of the conventional. He loved wrestling with machinery and defying danger. He once wrote to Rutherford, at that time traveling around the world, a typical letter about his experiments with a powerful new apparatus: "I am writing you this letter to Cairo to tell you that we already have a short-circuit machine and the coil and we managed to obtain fields of over 270,000 in a cylindrical volume of a diameter of 1 cm. and 4½ cm. high. We could not go further as the coil burst with a great bang which no doubt would amuse you very much if you could hear it. The power in the circuit was about 13½ thousand kilowatts . . . approximately three Cambridge Supply Stations connected together . . . the accident was the most interesting of all experiments . . . we know just what an arc of 13,000 amps is like. . . ."

Even when Kapitza was simply posing for an ordinary photograph, as he did in 1931 at the Physics Congress in Zürich, he had to be dramatic. He had himself photographed lying close to the wheels of a car, explaining: "I just want to know what I should look like if I were being run over."

Rutherford was tireless in securing new facilities for Kapitza's "baby giants" of high voltage. On the former's recommendation a special laboratory was built for them by the Royal Society and the Department for Scientific and Industrial Research, a government organization set up after the end of the First World War. Named after the chemist and multimillionaire Mond, this establishment was finally opened in February 1933. The astonished participants in the opening ceremony beheld, carved on the façade, an extraordinary heraldic beast. It was a crocodile which had been chiseled in the stone at the special request of Kapitza by the renowned British sculptor Eric Gill. When Kapitza was asked what so utterly outlandish a creature was doing there, he replied: "Well, mine is the crocodile of science. The crocodile cannot turn its head. Like science, it must always go forward with all-devouring jaws." "Crocodile," however, happened also to be the nickname invented by the Russian for Rutherford, as everyone in the Cavendish Laboratory knew except Rutherford himself.

Kapitza was not at first given the opportunity to use his new laboratory. In 1934 the Russian Academy of Science, when it moved from Leningrad to Moscow, elected him a member in spite of Stalin's opposition. Kapitza thereupon paid a visit to his native land. It was not the first time that this semiemigrant had undertaken a homeward journey. But this time things did not go quite so smoothly. He was told, when he was about to return to Cambridge, that the Soviet Union "could no longer dispense with his services, in view of the danger from Hitler." Kapitza thus became a prisoner in his own country. Rutherford wrote to Moscow requesting that he should be allowed, in the interest of science, to return to his duties. The Russian government replied: "Of course England would like to have Kapitza. We, for our part, would equally like to have Rutherford in the Soviet Union." After this cleverly phrased refusal of his request Rutherford appealed to the British Prime Minister, Baldwin: "Kapitza was commandeered as the Soviet authorities thought he was able to give important help to the electrical industry and they have not yet found out that they were misinformed." But even Baldwin's subsequent intervention produced no effect.

One of Kapitza's relatives, raising the question of his return with the Soviet ambassador in London, Maisky, concluded her appeal with the words: "You won't be able to keep him anyhow. Our Pyotr has a hard head." The diplomat is said to have assured the lady: "But our Joseph has a still harder one."

As all these efforts proved fruitless, Rutherford took a step which showed both his immeasurable faith in the international character of science and his affection for his lost favorite. He determined to send on to Kapitza in Russia the entire installation of his new laboratory, over the equipment of which he had taken so much trouble. The British scientists Adrian and Dirac traveled to Moscow to arrange the transfer of all this valuable and bulky apparatus. It was shipped aboard a Soviet freighter in an English harbor and soon afterwards arrived at Leningrad all mixed up with a cargo of frozen meat. The Russian government, to win Kapitza over, not only paid £30,000 for the dismantled Mond Laboratory but also built a brand-new institute for him in Moscow, in the style of an English gentleman's country seat. Kapitza resigned himself to his gilded cage. In 1936 he wrote to Rutherford: "After all, we are only small particles of floating matter in a stream which we call fate. All that we can manage is to deflect our tracks slightly and keep afloat—the stream governs us."

Kapitza's enforced departure not only affected Rutherford ery deeply. It also had a disrupting effect upon the Cavendish Laboratory as a whole, and during the next few years its magnificent team began to disintegrate. First Blackett went, then Chadwick and finally Oliphant. They accepted important appointments at other universities. Rutherford's own mighty frame, which had always been the very picture of health and strength, suddenly began to age, though he would never confess it. One day, while he was trying to insert a small, thin slip of gold leaf into his electroscope, his hands trembled so much that he had to ask his assistant, Crowe, to put it in for him. The same thing happened again a few days later. It worried Crowe, and he asked his chief: "Nerves not too good again today, sir?" Rutherford retorted in his dreaded leonine roar: "Nerves be damned! You're shaking the table!"

On October 14, 1937 the learnea scientist suffered a slight rupture after some strenuous work. A small operation, apparently perfectly safe, became necessary. But it turned out badly, and five days later the pioneer of experimental atomic research was dead. With him disappeared a scholar of the old breed whose desire to understand the nature of the world of atoms arose simply from his love of truth. When, as early as 1932, after the great successes of his team, the newspapers were already beginning to prophesy possible future applications of atomic energy to practical purposes, Rutherford immediately rebuked them. "It was not," he said, "that the experimenters were searching for a new source of power or the production of rare or costly elements. The real reason lay deeper and was bound up with the urge and fascination of a search into one of the deepest secrets of nature."

Feather, under Rutherford's direction at Cambridge, conducted some highly informative tests on the effects of neutrons, but the most interesting results of research after 1934 came from Rome. The Eternal City had for some years been developing into the capital of the world of physics, owing to the work of Enrico Fermi, then just over thirty. Fermi's decision to devote himself to atomic physics instead of, as before, to spectroscopy had been reached, quite casually, during a discussion with his team in the locker room of a tennis court. His subsequent success soon proved he was right. Even his first theoretical studies created a considerable sensation, especially among the younger generation of physicists. They made frequent journeys to Rome just to meet the Italian, who was to be taken seriously as a scientist despite all his boyish enthusiasm for sport.

Fermi did not disappoint them. A typical reaction was that of Bethe, Arnold Sommerfeld's star pupil, who wrote to his teacher from Rome: "Of course, I went to see and admire the Colosseum. But the best thing in Rome is undoubtedly Fermi. He has a marvelous faculty for immediately seeing the solution of any problem submitted to him." In 1934 when Curie's successful generation of artificial radioactive elements became known through his conclusive report to the Académie des Sciences, Fermi had just experienced a setback. His latest article on beta rays had been declined by *Nature,* the London periodical, then considered the most important of all those dealing with natural science. He meant accordingly to try his hand for once, "just for fun," as he said, at the sort of practical experiments Joliot had been undertaking. But Fermi, the "Pope," as he was called by his even younger collaborators, resolved to make use, not of the alpha rays which the French had been employing, but of the new, more powerful projectile, the neutron.

Fermi's able wife, Laura, has given a humorous account of how her husband and his pupils began in 1934, the "year of miracles," a systematic bombardment of one element after another with neutrons. Results with the first eight elements tested were negative. But with the ninth element, fluorine, the Geiger counter began to tick. Radioactivity had been artificially generated. The work was so exciting that d'Agostino, a young student of physical chemistry, who had originally only come from Paris to join the team as a guest for a few weeks, put off his departure for so long that his return ticket finally expired and he decided to stay for good.

In the course of these experiments Fermi and his closest collaborators made two important discoveries. The first was the extraordinary circumstance that the radioactivity of a metal bombarded with neutrons was a hundred times as great when the neutrons had previously been slowed down by water or paraffin—the conjecture had been first confirmed in the picturesque goldfish pond behind the Physics Building. Secondly, it was found that the bombardment of uranium, the heaviest of all metals, apparently gave rise to a new element, or perhaps even several new ones, the so-called artificial transuranic elements. The first discovery was later proved correct and had a decisive influence on the subsequent development of atomic physics. The second turned out to be an illusion.

The truth was that Fermi had not created new transuranic elements with his neutrons. He had—and probably he was the first to do so—split the uranium atom. Fermi's work, which seemed to have reached its climax with the production of a new element having the ordinal number 93, made a deep impression on the scientific world. He had been able to demonstrate the powerful effects of the neutron, the new microscopic particle discovered by Chadwick, even if he could not for the time being reveal the effects of neutron bombardment—in reality of far greater and more revolutionary significance.

Many laboratories now began to conduct experiments like Fermi's. Only one critical comment was heard in the midst of the general applause. At the Institute for Physical Chemistry of the University of Freiburg in Breisgau, Ida and Walter Noddack, a young couple engaged in research, had been keeping a lookout for natural transuranic elements ever since 1929. Frau Noddack had discovered in 1925, when she was still an adolescent, the hitherto unknown element of rhenium. She and her husband were regarded as the leading authorities on the chemical analysis of the "rare earths." In 1934 they received from a Czechoslovakian chemist named Koblic a specimen of a red salt he had extracted from the uranium mines in Joachimsthal. He believed it to be a transuranic element and wished to give it the name of "bohemium." The Noddacks proved conclusively, by chemical tests, that the Czech scientist's assumption was incorrect. Ida Noddack pronounced an equally unfavorable verdict on Fermi's "transuranic elements." She not only demonstrated that the Italian physicist had not really submitted in his chemical analysis any convincing evidence of his claim. She also advanced a bold hypothesis, which was first shown to have been justified at the end of 1938. For in 1934, more than four years before the discovery of the splitting of the uranium atom by Hahn and Strassmann, she wrote in the *Zeitschrift für angewandte Chemie* (Applied Chemistry Magazine): "It would be equally possible to assume that when a nucleus is demolished in this novel way by neutrons nuclear reactions occur which may differ considerably from those hitherto observed in the effects produced on atomic nuclei by proton and alpha rays. It would be conceivable that when heavy nuclei are bombarded with neutrons *the nuclei in question might break into a number of larger pieces* which would no doubt be isotopes of known elements but not neighbors of the elements subjected to radiation."

When this criticism and the suggestion arising out of it reached Fermi in Rome he did not take them seriously. To suppose that the neutrons, possessing a strength of less than one volt, might have split the atomic nucleus which was able to withstand bombardment of a strength of millions of volts, seemed to him, as a physicist, an impossible hypothesis. He was all the more convinced that he was right when Otto Hahn, the most famous expert on radium in the world, agreed with him.

At that time, in the laboratory of the Kaiser Wilhelm Institute in Berlin-Dahlem, Fermi's transuranic elements were being intensively studied under the direction of Hahn and his colleague from Vienna, Fräulein Lise Meitner. In numerous publications between 1935 and 1938 Hahn and Meitner described exhaustively the chemical properties of the new bodies which had come into existence as a result of bombardment by neutrons. Frau Noddack reports: "I had just as lasting doubts of their identification of the separate transuranic elements through the latter's chemical properties as I had about Fermi's interpretation. My husband and I had known Hahn well for decades and he had often inquired about the progress of our work . . . when in 1935 or 1936 my husband suggested to Hahn by word of mouth that he should at least make some reference, in his lectures and publications, to my criticism of Fermi's experiments, Hahn answered that he did not want to make me look ridiculous as my assumption of the bursting of the uranium nucleus into larger fragments was really absurd."

In order to understand why neither Fermi nor the Hahn-Meitner research team took Frau Noddack's hypothesis seriously, one must realize that according to the ideas of physics current at the time only projectiles of a penetrating power so far not attained would be capable of forcing a way into the nucleus of a heavy atom and splitting it. Since Rutherford's first experiments the artillery of the besiegers of the atomic nucleus had grown in strength and multiplicity. In the United States, especially, atom-smashing equipment, such as the Van de Graaff generators and cyclotrons, had been constructed. These machines were already capable of accelerating certain particles used as projectiles up to the enormous energy of nine million volts. Nevertheless, even they had only damaged, without breaking into, the protective walls with which Nature in her wisdom had encircled the atomic nucleus and the tremendous stores of energy it contained. The idea that neutrons, which carried no electrical charge at all, might have been able to accomplish what could not be done with such heavily charged projectiles was too fantastic to be credited. It was as though one were to suggest to troops which had been vainly shelling an underground shelter with guns of the heaviest caliber for a long time that they should start trying their luck with ping-pong balls.*

And yet it was not only on technical grounds that the atomic scientists of the period 1935-38 so often overlooked the truth. We may ask, for example, what Italy's war against Abyssinia meant to Fermi, who was just then making such splendid progress with his studies. We know from the statements of his collaborators how greatly the work of his team was disturbed by the war, by the turning of world opinion against Italy and, arising from this intensification of feeling, by the closer watch kept by politicians upon all intellectuals, including those of the Institute in the Via Panisperna. According to Segré the atlas in the Institute library soon started opening automatically at the page for Abyssinia. Instead of discussing bombardments with neutrons Fermi and his young men argued about the shelling of Abyssinian strongholds. The atmosphere was not one in which clear thinking and scientific self-criticism could thrive.

The accounts of research, in their scientific sobriety, make no mention of the influence of such political or even private interference. When they are read today the course of this comedy of errors becomes clear, with all its wrong turnings and gropings, which none of those concerned could ever understand later. The reports, though, say nothing whatever about the people who participated in the work, how they lived or what they felt.

The public remained unaware that between the two figures who played leading roles in the drama of the uranium experiments an increasing personal rivalry had developed, though nothing of the kind is ever admitted in polite scientific society. Madame Irène Joliot-Curie and Fräulein Lise Meitner were among the greatest experts in radium research of their time. No one disputed their pre-eminence. Yet a contest began between them, supported by their collaborators.

The friction started in October 1933 at the Solvay Congress

* An even better comparison, no doubt, would be that of the neutrons with "saboteurs" capable of making their way into the interior of the atom, not by force, but with the aid of a sort of cap of invisibility.

in Brussels. Madame Joliot had given an account, with her husband, of her bombardment of aluminum with neutrons. What happened next is told by Joliot: "Our report led to a lively discussion. Fräulein Meitner announced that she had made similar experiments but had not obtained the same results. In the end the great majority of the physicists present came to the conclusion that our experiments had been inexact. After the session we felt rather desperate. At that moment Professor Bohr approached us and said he considered our data extremely important. Shortly afterwards Pauli encourged us in similar fashion."

The Joliots, on their return to Paris, resumed their work. The report which Fräulein Meitner had criticized as inaccurate in Brussels became the foundation of the Joliots' most important discovery—artificial radioactivity. This did not improve relations between the laboratories in the rue d'Ulm and Berlin-Dahlem. Hahn even complained to Rutherford of the Joliots later on. The British scientist answered: "I am sorry that I have unwittingly upset you by my reference to the thorium-neutron transformations. Actually, I wrote my paper when I was in my cottage in Wiltshire on holiday, where I had no papers to refer to. I had glanced through the Joliots' paper and thought that the evidence was rather vague and I gave expression to that view in my address. I had for the moment forgotten that you had actually published your letter in *Naturwissenschaften* before their paper appeared and I quite agree that they ought to have specifically mentioned your definite statement about the $4n + 1$ series. . . . I shall make a point of putting the matter right when I have the opportunity."*

It was due to such animosities that in Dahlem the studies

* After Madame Joliot-Curie's death in 1956—from leukemia, which she had contracted as a result of her work with radioactive matter—Fräulein Meitner composed a glowing obituary in memory of her colleague, in which she sought to elucidate the difficulties which had arisen between them thirty years before. She wrote: "She seemed to be afraid of being regarded rather as the daughter of her mother than as a scientist on her own account. This fear may have influenced her attitude to strangers. She was also entirely indifferent to social conventions. She had a strong inner feeling of self-sufficiency which might be mistaken for a lack of amiability."

published by Madame Joliot-Curie were much too casually designated as "unreliable." On one occasion, in 1935, Fräulein Meitner directed her pupil von Droste to repeat in Dahlem certain experiments which had been carried out in Paris with the bombardment of thorium. Madame Joliot-Curie had declared that the thorium isotope sent out alpha rays under radiation. Droste did not find any such rays. Once more Fräulein Meitner believed she had convicted her rival of inaccuracy. And once more she was mistaken.

Droste experimented not only with thorium, but also with uranium. If he had not, in the latter case, introduced a filter to avert particles with a range of under three centimeters, he would not only on that occasion have realized that Madame Joliot's results were as she had stated but would also have necessarily found, there and then, fission products from uranium. So near had experiment come, even at that date, to the discovery of uranium fission.

Irène Joliot-Curie's next paper on transuranic elements appeared in the summer of 1938, with the Yugoslav Savitch as coauthor. They mentioned a substance which did not fit in with the pattern for these elements worked out meanwhile by Hahn and his collaborators.

It was said in Dahlem, "Madame Joliot-Curie is still relying on the chemical knowledge she received from her famous mother and that knowledge is just a bit out of date today." Hahn considered it necessary to be tactful. He thought it desirable not to reveal his French colleague's negligence to the entire world in a scientific periodical. "There is quite enough vexatious strain just now in the relations between Germany and France," he said. "Let us not help to increase it." He accordingly wrote a private note to the laboratory in the rue d'Ulm suggesting that the experiments should be repeated somewhat more carefully.

But no reply to Hahn's note came from Paris. On the contrary, Madame Joliot committed a further "sin." She published a second article, based on the data of the first. Hahn refused to read it, in spite of being urged to do so by his assistant Strassmann. He was thoroughly disgusted with the unteachability of his Parisian colleague. Besides, he was being worried, that same summer of 1938, by a problem which had nothing to do with physics. Attempts were being made to deprive him of his alter ego. For over a quarter of a century Lise Meitner and her "cockerel"* had worked side by side. Their identities were so closely fused, even in their own minds, that Fräulein Meitner once absent-mindedly replied to a colleague who spoke to her at a congress: "I think you've mistaken me for Professor Hahn."

As an Austrian, despite the sudden discovery that she was "not Aryan," Lise Meitner had been permitted to go on working at the Kaiser Wilhelm Institute after 1932. The *Anschluss* of March 1938 made the racial legislation of the Third Reich applicable to her case. Intervention by Hahn and Max Planck, who even went to Hitler himself, proved useless in saving their colleague for the Institute. She had to go. It was not even certain whether the government would allow her to leave Germany, so she was forced to slip across the Dutch frontier, disguised as a tourist, without saying good-by to her former associates. Apart from Hahn, only two or three people in the Dahlem Institute knew that Lise Meitner would not come back from her summer holiday.

That autumn Madame Joliot-Curie published a third article which summarized and enlarged her previous two. Since Fräulein Meitner's departure Strassmann had become Hahn's closest associate in the field of radium chemistry. He saw at a glance that no mistake had been made in the Curie laboratory but that on the contrary a remarkable new avenue of approach to the problem had probably been opened up. He rushed excitedly upstairs to Hahn and exclaimed with the greatest emphasis: "You simply must read this report!"

Hahn was adamant. "I'm not interested in our lady friend's latest writings," he answered, puffing quietly at his cigar.

Strassmann would not give in to this rebuff. Before Hahn could repeat it he gave his chief a succinct account of the most important points in this new performance. "It struck Hahn like

* *Translator's note: Hahn* means "cock" in German.

a thunderbolt," he recollected later. "He never finished that cigar. He laid it, still glowing, on his desk, and ran downstairs with me to the laboratory."

It proved very difficult to persuade Hahn that he, in common with investigators all over the world, had been following a false trail for years; but as soon as he had recognized the fact he instantly turned in his tracks and made every effort to get at the truth. It was not easy to confess to a series of failures. To that admission, however, he owed, shortly afterwards, the greatest success of his career.

In almost incessant work, for weeks on end, the experiments of Madame Joliot and Savitch were thoroughly tested with the most exact methods of radium chemistry. The process showed that the bombardment of uranium with neutrons did in fact produce a substance which, as the Paris team had stated, closely resembled lanthanum. The more precise analyses of Hahn and Strassmann, however, led to the chemically incontrovertible but physically inexplicable result that the real element concerned was barium, which occupies a position in the center of the list and weighs a little more than half as much as uranium.*

The discovery was made only later that this at first incomprehensible presence of barium could be explained by the "bursting," as Hahn called it, of the nucleus. At the time what Hahn and Strassmann had found in their chemical probings seemed so incredible to them that they jotted down these skeptical sentences, which have since become famous:

"We come to this conclusion. Our 'radium' isotopes have the properties of barium. As chemists, we are in fact bound to affirm that the new bodies are not radium but barium. For there is no question of elements other than radium or barium being present. . . . As nuclear chemists we cannot decide to take

* Hahn himself wrote in this connection to the author: "The Paris people never mentioned barium but only lanthanum. They found that their mysterious substance, to which they had formerly ascribed different properties, closely resembled lanthanum, to such an extent, indeed, that they could only separate it from the latter by means of fractional crystallization. Such was the decisive error which prevented Joliot and Savitch from hitting upon the means of splitting uranium."

this step in contradiction to all previous experience in nuclear physics."

The two German atomic scientists perceived that they had made a notable discovery, even though it might not yet be explicable in terms of physics. The date was just before Christmas 1938 and it seemed important to Hahn to publish an account of his work as soon as possible. He took the unusual step of telephoning to Dr. Paul Rosbaud, the Director of Springer Verlag, a personal friend, asking whether he could find room for some urgent information in the forthcoming number of *Naturwissenschaften*. Rosbaud agreed to do so.

The paper, dated December 22, 1938, accordingly left Hahn's desk. Nearly twenty years later Hahn told the author: "After the manuscript had been mailed, the whole thing once more seemed so improbable to me that I wished I could get the document back out of the mail box."

It was with such hesitations and doubts that the age of atomic fission began.

As a demonstration against the racial legislation of the Third Reich, and yet at the same time no more than plain proof of a confidence that had now lasted for decades, Otto Hahn immediately dispatched his new data to his former collaborator, Fräulein Meitner, now an emigrant in Stockholm. The letter was already on its way before any other member of Hahn's department in the Kaiser Wilhelm Institute heard anything about the new and still, for the present, quite inexplicable discovery. Hahn waited tensely to hear what his former partner would say to these astonishing new results, which contradicted all previous experience. He was a little afraid of what her answer might be. She had always been a stern critic of his work. Probably, he thought, she would tear these latest data to pieces.

Lise Meitner received his letter in the small township of Kungelv, near Göteborg. She had gone to this seaside resort, almost lifeless in winter, to spend her first Christmas in exile alone in a small family boardinghouse. Her young nephew, the physicist O. R. Frisch, since 1934 himself a refugee working at the Institute of Niels Bohr in Copenhagen, felt sorry for his aunt's loneliness. He determined to pay her a visit. He was there when Hahn's letter arrived in the quiet little provincial town. At once it greatly excited his aunt. If the radium chemistry analyses carried out by Hahn and Strassmann were accurate—and Fräulein Meitner, who knew the precision of Hahn's work, could hardly doubt it—then certain ideas in nuclear physics, hitherto taken to be unassailable, could not be true. She perceived even more clearly than had Hahn that something tremendous had unexpectedly come to light.

Fräulein Meitner could hardly wait to discuss the multitude of questions and conjectures that arose in her mind. It seemed a bit of luck that her nephew, regarded as one of the leading lights in Bohr's circle, happened to be with her. But Frisch had not come to Kungelv to "talk shop" with his aunt; he had come for a holiday. "It took her some time before she could make me listen," he reported later. To be precise, Frisch actually tried to run away from Lise Meitner's explanations. He buckled on a pair of skis and would certainly soon have been beyond the reach of his aunt if the ground had not been so hopelessly flat all round the town. She was able to keep pace with him and maintained a continuous flow of talk while they stamped through the snow. At last, by a bombardment of words, she pierced the blank wall of his indifference and released a chain reaction of ideas in his brain.

Later that evening and the following days of a "spoiled holiday," inspiring debates took place in the old-fashioned lounge of the boardinghouse. Frisch described them in these words:

> Very gradually we realized that the break-up of a uranium nucleus into two almost equal parts . . . had to be pictured in quite a different way. The picture is one . . . of the gradual deformation of the original uranium nucleus, its elongation, formation of a waist and finally separation of the two halves. The striking similarity of that picture with the process of fission by which bacteria multiply caused us to use the phrase "nuclear fission"* in our first publication.

* It was the American biologist James Arnold, also at that time working with Bohr in Copenhagen, who suggested this technical phrase from

That publication was somewhat laboriously composed by long-distance telephone (Professor Meitner had gone to Stockholm, and I had returned to Copenhagen) and eventually appeared in *Nature* in February 1939. . . . The most striking feature of this novel form of nuclear reaction was the large energy liberated. . . . But the really important question was whether neutrons were liberated in the process, and that was a point which I, for one, completely missed.

Frisch was then still somewhat uneasy about his discovery. He wrote to his mother: "Feel as if I had caught an elephant by its tail, without meaning to, while walking through a jungle. And now I don't know what to do with it."

The news of Hahn's discovery and its momentous significance for physics, as shown by Fräulein Meitner and Frisch, at first caused mainly perplexity among atomic scientists. When Frisch, on his return from Sweden, gave an account in Copenhagen of Hahn's work and of his own conversations with his aunt Bohr struck himself a blow on the forehead.

"How could we have overlooked that so long?" he cried.

Arnold's own branch of science when Frisch described the phenomena to him.

Five

The Decay of Confidence (1939)

For three hundred years most new discoveries which threw light on the darkness of Nature had been welcomed in the name of progress. But in January 1939 many scientists hesitated in alarm. The fear of war, like a heavy cloud, hung over the world. Only the surrender of the democracies at the Munich conference had once more preserved international peace.

This sacrifice at Munich had done nothing to ease the tension. At that very moment a small group of initiated persons became aware of a source of power wholly novel, of superhuman dimensions. It was still possible, even then, however, to shade one's eyes, with the dark glasses of skepticism, from the dazzling and terrifying prospect of released atomic energy. As late as the beginning of 1939 Niels Bohr specified to his colleague Wigner, who had been working for ten years at Princeton, fifteen weighty reasons why, in his opinion, practical exploitation of the fission process would be improbable. Einstein assured the American reporter W. L. Laurence that he did not believe in the release of atomic power. And Otto Hahn, according to the young German physicist Korsching, exclaimed during a discussion with a few of his closest colleagues on the practical exploitation of his discovery: "That would surely be contrary to God's will!"

Up to that time all nuclear-fission experiments had been undertaken with such tiny quantities of uranium that there could be no question of the development of any considerable degree of power. The hopes and fears even then beginning to arise in the minds of atomic scientists could take shape only if it proved possible to increase to an enormous extent, as by the gathering force of an avalanche, the present miniature scale of the effects of splitting the atom. A chain reaction of this kind had been described as a theoretical possibility, as early as 1932–35, by Houtermans, Szilard and Joliot-Curie. But it could be achieved in practice only if, after fission of a nucleus of uranium, there were invariably released a number of supplementary additional neutrons which would then split further nuclei. Until this all-important question had been thoroughly examined and its validity confirmed, most atomic physicists assured colleagues already wondering what would then happen that there could be no real grounds for anxiety.

Parallel to investigations, half in hope and half in fear, of the possibility of a chain reaction, was an equally unusual political experiment, amounting to a new chapter in the history of ideas. Its author was Szilard, who had emigrated from England to the United States. As soon as he had heard from Bohr and Wigner about the experiments being conducted in Dahlem and Copenhagen, Szilard had his own experimental apparatus, left behind at Oxford, sent after him. He borrowed two thousand dollars from a friend named Liebowitz, a small New York manufacturer, as security for a gram of radium. Szilard had not yet received any university appointment in the United States, but he was granted special permission to work in the physics laboratory of Columbia University. After three days his first experiments seemed to indicate the possibility of an emission of additional neutrons. Szilard became more anxious than ever about the inevitable consequences of such experiments; no doubt they were also being carried out in Europe. His vivid imagination again outstripped events and he perceived with shocking clarity a possible race in the production of atomic armaments.

Something had to be done. One of the first people consulted by Szilard was Fermi. The Italian scientist had started on his journey to Stockholm, where the Nobel prize awaited him, in November 1938, with the firm intention of not returning to

Fascist Italy. He was now working in the same university building as Szilard, and he and the young American Herbert Anderson were also studying the problem of neutron emission. Fermi at first unhesitatingly rejected the idea suggested to him by his Hungarian colleague that scientists should themselves impose a voluntary censorship on their work. He had, after all, just escaped from a country where censorship and secrecy regulations had crippled intellectual activity.

The reactions of most of Szilard's other colleagues to his proposal were not much more encouraging. At first only three of them agreed with him: Wigner, Teller (who had been summoned in 1935, on the recommendation of his friend Gamow, to George Washington University in the American capital) and Weisskopf, just over from Copenhagen to accept an invitation to the University of Rochester.

These four men were not impressed by the counterargument that science had fought for centuries for the free exchange of ideas and should never lend its support to the opposite principle. They had been themselves, throughout their lives, devoted adherents of the maxims of liberty and uncompromising antagonists of militarism. It now seemed to them, nevertheless, that a highly exceptional situation existed.

The world was asking itself how Hitler could afford to challenge the Great Powers. If his sources of raw materials and his production capacity were compared with his political adversaries' throughout the world, it was obvious that the Führer, despite his temporary superiority in aircraft and tanks, could never hope to be victorious. Or was there, perhaps, some unknown factor which might upset the confident calculations of the Allied Powers? The few physicists who were already aware at that period of the terrifying possibility of an atom bomb felt that they were justified in suspecting that it was the prospect of the employment of this extraordinary new weapon which constituted the unknown quantity, the "x," in the 1939 power-politics equation. If this conjecture were correct, it could not only explain the steadily growing recklessness of Hitler's provocations, but might also actually mean that he was quite consciously heading for a war he possibly intended to win by playing the ace of the uranium bomb. If he alone possessed atom bombs he would be unconquerable, despite his economic weaknesses. The German dictator would then be in a position to enslave the entire world. What could be done to prevent his achieving this aim?

Why did not the atomic-research scientists in the United States first attempt to discuss this fateful question with their colleagues in Germany? Mutual trust within the family of atomic physicists was already too much undermined to permit it. It was known outside Germany, of course, that Hitler's government was on bad terms with physicists. The regime had been openly criticized by the Nobel prize winner von Laue; he remained in the country only because he believed that the few teaching appointments available abroad should be reserved for those whom Hitler had forced into exile. It was also known that Heisenberg and all the adherents of up-to-date physics had been attacked in 1937 by the *Schwarze Korps* as "white Jews." * It was known that a miniature war had broken out openly at the Würzburg Physics Congress in 1934 between the faction of the so-called "German physicists" and those of their colleagues who maintained that neither "German" nor "Jewish" physics existed, that physics was either true or false. Nevertheless, physicists outside Germany considered this resistance by reasonable men of science too feeble to guarantee a possible secret pact between scientists whose aim it would have been to call a halt to the further development of atomic research. Above all, one could never be certain that German physicists, so long as they remained in Hitler's power, would not simply be compelled by blackmail or the use of brute force to serve the ends of National Socialism. Such was the opinion of the American physicist P. W. Bridgman. He announced in the periodical *Science*, at the end of February 1939, that he intended henceforth, with regret, to forbid scientists from totalitarian countries access to his

* *Das Schwarze Korps* was the official weekly of the SS, aggressive in tone and greatly feared.

laboratory, and that he expected his colleagues to take similar steps. He wrote:

> I have had the following statement printed, which I hand to any prospective visitor who may present himself. "I have decided from now on not to show my apparatus to or discuss my experiments with the citizens of any totalitarian state. A citizen of such a state is no longer a free individual, but may be compelled to engage in any activity whatever to advance the purposes of that state. . . . Cessation of scientific intercourse with totalitarian states serves the double purpose of making more difficult the issues of scientific information by these states and of giving the individual opportunity to express abhorrence of their practices."

Amazingly few voices were raised at that time against this break with the traditions of science. The conviction was almost universal that the new, unprecedented methods of the dictators in contravention of the principles of humanity and liberty—methods which had only recently been demonstrated afresh by the occupation of Prague—must be countered by equally unprecedented measures.

It thus happened that in the United States Szilard's idea gradually came to prevail. The self-imposed censorship of scientists was of course only to operate against supporters of the Axis Powers. Even Fermi, who had previously declared that he would have nothing to do with it, now agreed to the voluntary self-censorship.

Szilard's group found it more difficult to obtain the consent of European atomic-research scientists to their unconventional idea of keeping all further work in nuclear physics a secret. Szilard had written to Joliot-Curie as early as February 2, 1939, to prepare him for the measure in view: "When Hahn's paper reached this country about a fortnight ago, a few of us at once got interested in the question whether neutrons are liberated in the disintegration of uranium. Obviously, if more than one neutron were liberated, a sort of chain reaction would be possible. In certain circumstances this might then lead to the construction of bombs which would be extremely dangerous in general and particularly in the hands of certain governments."

Szilard's letter ended with an observation which showed the extent to which previous hopes of progress by scientists had degenerated, under the influence of the sinister possibilities of this development, into actual fear of progress: "We all hope that there will be no or at least not sufficient neutron emission and therefore nothing to worry about." It was as though he had expressed a wish for the experiments to fail.

Szilard had told Joliot that he would send him a cable if agreement were reached to proceed with the voluntary withholding of research data. He had also requested to be informed in general terms of his correspondent's attitude, but he never heard from him about it. There was a very good reason for Joliot's silence. With his collaborators Hans von Halban and Lew Kowarski, Joliot was just on the point of experimental realization of the chain reaction to which Szilard's anxious communication had referred. He was determined not to be deprived, under any circumstances, of the credit for being the first with this discovery. When the experiment succeeded a month later, he did not entrust the account of it, as in the case of all his previous work, to a French periodical. He sent his report to the British magazine *Nature* because it usually published the work sent in to it more quickly than any other journal concerned with natural science. To make assurance doubly sure that this important communication should arrive in London without fail in time for the next issue, Kowarski traveled, on March 8, to the airport of Le Bourget, only an hour's journey from the center of Paris, and personally supervised the document's deposit in the London mailbag. To such a race, for the sake of a few days, had atomic research already degenerated by the spring of 1939. A wholly new spirit of keen international competition had now arisen.

As soon as Szilard realized that Joliot had apparently taken no notice of his letter, he and his friends redoubled their efforts, so as to stop the publication of any further studies. The British had been waiting to see what would come of the steps taken in the United States before committing themselves. Now they pledged their support to the movement. John Cockcroft, a former member of Rutherford's team, answered a letter from Wigner in the middle of April. He was still skeptical of the success of the scheme but said he was resolutely in favor of it. He wrote: "Dirac has given me your message about uranium. Up to the present I have felt that it is very unlikely that anything usable can come out of this in the next few years. However, under the present circumstances we cannot afford to take any chances."

On the other hand, Joliot-Curie still seemed indifferent to the whole affair. Not until Weisskopf had sent him a telegram of a hundred and fifty words, emphasizing the seriousness of the matter, did he reply from Paris, at last, by cable:

RECEIVED LETTER SZILARD BUT NOT CABLE PROMISED STOP PROPOSITION OF 31 MARCH VERY REASONABLE BUT COMES TOO LATE STOP HAVE LEARNED LAST WEEK THAT SCIENCE SERVICE HAD INFORMED AMERICAN PRESS FEBRUARY ABOUT ROBERTS WORK STOP LETTER FOLLOWS JOLIOT HALBAN KOWARSKI.

It was quite true, as Joliot stated, that some information had been given to the press. The release, however, was in such general terms that as a ground for their abstention from the movement the resort to it could only be regarded as prevarication. As a matter of fact Joliot-Curie had been influenced by a number of considerations. In the first place he had not taken Szilard's letter seriously simply because he had supposed it to be a solo performance by his Hungarian colleague. Weisskopf's telegram, which happened to arrive on the April 1, All Fools' Day, in Paris, had strengthened this impression that the proposal had been made unofficially by a minority of scientists. So important a matter, in the opinion of the formally minded French, should have been broached by the American Academy of Sciences instead of being raised by a few "individualists" and "outsiders."

Stronger than this psychological argument was still another reason about which one of the three members of Joliot's team expressed himself freely: "We knew in advance that our discovery would be hailed in the press as a victory for French research and in those days we needed publicity at any cost, if we were to obtain more generous support for our future work from the government."

The publication of Joliot's views so increased the resentment among Szilard's American colleagues against the censorship to be imposed by themselves on their own work—a censorship to which in any case they had only agreed with reluctance—that Professor Rabi went to see Szilard and warned him that if he did not give way in this matter he could no longer count on the hospitality of Columbia University, where he had hitherto only been allowed to work as a guest. Szilard had to acquiesce, under protest, in the future publication of his own pioneering investigations into chain reactions in uranium.

In the course of the controversy Wigner made a suggestion destined to have important consequences. He proposed that the American government should be informed of the "uranium situation." He argued that this step was necessary to enable the authorities to meet "any sudden threat" which might arise—meaning the development by Hitler of an atom bomb.

From the end of April until the end of July 1939 Szilard and his friends were deeply concerned with the question of how best to impress upon the American government the importance of the latest researches in atomic science and their possible effects upon the technique of war. The first attempt to interest a public authority in this matter had, to all appearance, utterly failed. On March 17, 1939, Fermi called on Admiral S. C. Hooper, Director of the Technical Division, Naval Operations, with a letter of introduction from Dean George Pegram of Columbia. The possibility of an atom bomb was discussed. But apparently the information offered by Fermi on this subject made no great impact upon the Admiral. At any rate, for the time being neither Fermi nor any other atomic scientist was invited to any further debate on the question. Curiously enough not even a report which appeared in the New York *Times* at the end of April, dealing with the spring meeting of the American Physical Society, aroused any interest among the authorities in Washington. And yet, no less a person than Bohr affirmed on that occasion in public that a bomb containing a minute quantity of uranium 235 bombarded by slow neutrons would be bound to set off an explosion powerful enough to blow up, at the lowest estimate, the entire laboratory and most of the surrounding town.

Szilard, Wigner, Teller and Weisskopf had to overcome both internal and external inhibitions before they could achieve contact with the American government. In the first place, as former Central Europeans, they felt very little confidence indeed, as a matter of principle, in any government, and least of all in military authorities. Secondly, not one of them was a native American. With the exception of Wigner, none of them had even been immigrants long enough to have obtained citizenship.

While Szilard and his friends were still racking their brains to find means of attracting the attention of some really influential authority, they received confidential information that work was already progressing on the uranium problem in the Third Reich, with the knowledge and support of the German government. This news seemed to confirm their worst fears.

The report was true. In April 1939 Dr. Dames, head of the Department of Research in the *Ministerium für Wissenschaft, Erziehung und Volksbildung* (Ministry of Science, Education and National Culture), had received a communication from two physicists, Joos and Hanle, mentioning the possibility of a "uranium machine." He called a conference at Berlin for April 30, attended by six German experts in atomic science. Hahn, the discoverer of nuclear fission, was not among the elect. His omission was formally justified on the ground that he was not a physicist but a chemist. But the true reason was that the top men knew perfectly well that Hahn was no friend of Nazism, and in scientific circles he was even said to have exclaimed: "I only hope you physicists will never construct a uranium bomb. If Hitler ever gets a weapon like that I'll commit suicide."

At the first meeting in the building at No. 69 Unter den Linden nothing at all was in fact said, for the time being, about an atom weapon. Those present simply discussed the possibility of employing nuclear fission to drive motor vehicles. After a survey by Joos of the position of atomic research, both abroad and at home, it was resolved to carry out an experiment of this kind. It was to be conducted as a joint undertaking by the leading physicists active in the field concerned. Those who participated in this session of April 30 were directed to keep the matter secret. But one of them, Mattauch, did not obey this order. That same evening he reported the discussions at the Ministry to Dr. S. Flügge, one of Hahn's closest and most gifted collaborators. Flügge's reaction was the exact opposite of those of his colleagues across the Atlantic. It seemed to him that immense risks would be run if a scientific discovery that might have such tremendous political consequences were kept out of the public view. He personally had not been sworn to secrecy. Consequently he wrote a detailed account of uranium chain reaction for the July number of *Naturwissenschaften.* This was followed, as further elucidation, by an interview granted by Flügge himself, at his own suggestion, in more generally intelligible terms, on the same subject, to a representative of the conservative *Deutsche Allgemeine Zeitung,* a newspaper which the Nazis only tolerated with reluctance.

Unfortunately Dr. Flügge's publicity only increased the alarm in America. People could not imagine even a line of print appearing in Germany without the express intention or consent of the government: "If the Nazis allow as much as that to be published about the uranium problem," it was considered in the United States, "they are certain to know a lot more about it. Consequently, we had better hurry. . . ."

It was a false deduction. In the summer of 1939 one more chance, quite unexpected, arose for personal contact with the German atomic scientists. Heisenberg was then on a visit to the United States. Pegram, chairman of the Physics Department of Columbia University, was aware of Szilard's and Fermi's efforts. He tried to induce the German physicist, by the offer of a professorship, to remain in America, but Heisenberg declined the offer so he would not have to desert the "nice young physicists" —such was the expression he used—entrusted to his charge back in Germany. He was sure, he added, that Hitler would lose the war. However, he wished to be in Germany during the coming catastrophe to help preserve what was valuable in his country.

A further cautious attempt was made by Fermi at Ann Arbor, where he was lecturing that summer at the University of Michigan. When he met Heisenberg at the house of their Dutch colleague, Samuel Goudsmit, their conversation inevitably turned to the fascinating problems raised by Hahn's discovery.

Heisenberg said on a later occasion: "In the summer of 1939 twelve people might still have been able, by coming to mutual agreement, to prevent the construction of atom bombs." He himself and Fermi, who were undoubtedly included among the twelve, ought then to have taken the initiative. But they let the opportunity go by. Their powers of political and moral imagination failed them at that moment as disastrously as did their loyalty to the international tradition of science. They never succeeded in achieving thought and action appropriate to the future consequences of their invention. Nor had they, in that critical situation, enough confidence in the legacies bequeathed by the past of their profession.

"The fact that we physicists formed one family was not enough," Weizsäcker remarked after the war. "Perhaps we ought to have been an international Order with disciplinary power over its members. But is such a thing really at all practicable in view of the nature of modern science?"

Six

The Strategy of Prevention (1939-1942)

That summer the news which reached the United States about the progress made by the German uranium project became more and more alarming. In Berlin there had been a second meeting of nuclear physicists. This time it had been called by the head of the Research Division of the Army Weapons Department. Colonel Schumann had acted on the basis of information from the Hamburg physicist Harteck. This researcher had drawn attention at the end of April to the "possibility, in principle, of the release of a chain reaction in uranium." According to a later report by his associate, Diebner, he had recommended its investigation by the German War Office.

A further item of news which reached the physicists in America by secret channels seemed to indicate that the Germans were really in earnest. They had suddenly forbidden all exports of uranium ore from Czechoslovakia, which they had recently occupied.

The only other place in Europe with any large stocks of uranium was Belgium, which obtained it from the Belgian Congo. Should not something be done at once, thought Szilard, to safeguard from seizure by Hitler this metal which had now become strategically so important? The American State Department did not yet realize that uranium might have any military significance at all—this rare metal had been used almost entirely for the production of luminous figures on dials and for the manufacture of pottery.

It was then that it occurred to Szilard for the first time that Einstein might help. Einstein belonged to the small international group of friends, outstanding intellectually and musically, which the Belgian Queen Mother, Elisabeth, had gathered around her in the course of her life. It might perhaps be possible to convey a warning to the Brussels government through this connection. An appointment was soon made with the father of the theory of relativity, who lived at Princeton, where Szilard's close associate Wigner also lived. Einstein was about to leave for his summer holiday on Long Island, but he had no objection to his two colleagues paying him a visit there to discuss their important project.

Accordingly, Wigner and Szilard, one hot day in July 1939, set off for Patchogue on the south coast of Long Island. After a drive of two hours, they discovered that apparently they had been given the wrong address.

"Perhaps I misunderstood the name Patchogue on the telephone," Wigner said. "Let's see if we can find some similar name on the map."

"Could it be Peconic?" asked Szilard after some minutes of strained silence.

"Yes, that was it," Wigner answered. "Now I remember!"

At Peconic the two travelers made exhaustive inquiries as to the whereabouts of the cabin of a Dr. Moore, whose small house Einstein had rented. A group of vacationers in shorts and brightly colored bathing suits came sauntering along. "No, we don't know Dr. Moore's cabin," they answered. Nor did the local inhabitants seem to have any information on the subject.

The two men continued to drive around, though it seemed a hopeless venture. Suddenly Szilard exclaimed: "Let's give it up and go home. Perhaps fate never intended it. We should probably be making a frightful mistake by enlisting Einstein's help in applying to any public authorities in a matter like this. Once a government gets hold of something it never lets go. . . ."

"But it's our duty to take this step," Wigner objected. "It must be our contribution to the prevention of a terrible calamity." So the pair continued their search.

"How would it be," Szilard proposed eventually, "if we simply asked where around here Einstein lives? After all, every child knows him."

The idea was immediately put to a practical test. A sunburned little boy about seven years old was standing on a street corner absorbed in adjusting his fishing rod.

"Do you know where Einstein lives?" Szilard asked him, more by way of a joke than in earnest.

"Of course I do," retorted the youngster. "I can take you there if you like."

The visitors had to wait for a short time on the open veranda before Einstein came out, wearing slippers, and escorted them to his study. Szilard's account of this first important conversation runs as follows:

> The possibility of a chain reaction in uranium had not occurred to Einstein. But almost as soon as I began to tell him about it he realized what the consequences might be and immediately signified his readiness to help us and if necessary "stick his neck out," as the saying goes. But it seemed desirable, before approaching the Belgian government, to inform the State Department at Washington of the step contemplated. Wigner proposed that we should draft a letter to the Belgian government and send a copy of the draft to the State Department, giving it a time limit of a fortnight in which to enter a protest if it believed that Einstein should abstain from any such communication. Such was the position when Wigner and I left Einstein's place on Long Island.

Szilard was now again confronted by the same obstacle which had been occupying his mind for weeks. How could he make sure of the attention of the American government? He discussed the problem with a number of friends, including Gustav Stolper, the German economist and former editor of the periodical *Der deutsche Volkswirt* (*German Economist*), who had emigrated to New York. Stolper had an idea. He was acquainted with a man who was known, though he was not himself a government official, to have the ear of President Roosevelt. This was the banker and scholar Alexander Sachs. This international financier could always obtain entry to the White House, for he had often amazed Roosevelt by his usually astonishingly accurate forecasts of economic events. Ever since 1933 Sachs had been one of the unofficial but extremely influential advisers of the American President, all of whom had to possess, by F.D.R.'s own definition, "great ability, physical vitality and a real passion for anonymity."

Sachs at once supported Szilard's ideas enthusiastically. During the next two weeks the two men drafted, in Sachs's office in the Wall Street investment firm of Lehman Brothers, a letter which went further in its terms than the document Einstein had originally been willing to sign. It was now intended that the draft should go not, as formerly planned, to the State Department, but to the White House. More rapid and energetic action was to be expected from the President than from the Secretary of State. The draft dealt with the point discussed with Einstein—the need for American negotiations with the Belgian government on the subject of the safeguarding of the stocks of uranium from the Congo. But in particular, a second point now added was a proposal for the financial support and acceleration of atomic research. The authors of the letter deliberately refrained from asking for any government assistance, but merely suggested that a confidential agent of the White House should be appointed to obtain the co-operation of private individuals and industrial laboratories in the secret project envisaged.

On August 2 Szilard again drove out to Long Island. Wigner had by that time gone to California, so Szilard's driver was his young countryman Edward Teller, later to play a further important part in the fateful drama lived by the atomic scientists. Did Szilard already have the final text of the draft in his pocket that day? Both Teller and Einstein state that he had. Einstein always declared that he merely signed this document. Szilard, on the other hand, observes: "So far as I remember, Einstein dictated a letter to Teller in German and I used the text of that letter as a basis for two more drafts, one comparatively short and one rather longer, both addressed to the President. I left it to Einstein to decide which he preferred. He chose the longer draft. I also prepared a memorandum as an enclosure to Einstein's letter. Both letter and memorandum were handed to the President by Dr. Sachs in October 1939."

This version of what took place seems the more probable to Dr. Otto Nathan, who had known Einstein for many years and later acted as his executor. Teller, however, affirms positively: "Einstein only signed his name. I believe that at that time he had no very clear idea of what we were doing in nuclear physics." Sachs, too, states, not without a certain degree of cynicism: "We really only needed Einstein in order to provide Szilard with a halo, as he was then almost unknown in the United States. His entire role was really limited to that."

"I really only acted as a mail box. They brought me a finished letter and I simply signed it." Such was the apology offered by Einstein after the Second World War to Antonina Vallentin, his old friend and biographer. He very soon began to regret his action. This gifted man of learning and great friend of peace explains further, clearly, in personal letters and notes which will probably be made public in future years how through a paradox of fate he had decided to give the starting signal for the most horrible of all weapons of destruction.

Einstein was of course at that time convinced that the government, which he had recommended take an active interest in the uranium problem in order to guard against possible surprise by a German atom bomb, would handle the tremendous new power entrusted to it with true wisdom and humanity. He acted on the assumption that the United States would never use such a bomb for any object other than self-defense against a similar weapon and even then only if its own safety were imperiled to an extreme degree. But when six years later the first atom bomb was employed against Japan, a country already on the verge of capitulation, he felt that both he himself and the atomic scientists who had worked on the construction of the weapon had been deceived.

The tragedy of the decision taken by the pacific-minded Einstein deepens when one realizes, as is possible today, that the menace of a German uranium bomb, doubtless believed perfectly genuine by the eminent scholar and those who influenced him, was in fact nothing more than a terrible phantom.

Einstein said, with deep regret, after the war: "If I had known that the Germans would not succeed in constructing the atom bomb, I would never have lifted a finger." The ability of the Third Reich to produce new weapons capable of deciding the issues of war was then greatly overestimated throughout the world. Subsequent investigations by Allied committees proved that on the outbreak of war the German leaders erroneously believed that they could achieve final victory with the weapons they already had. It was not until 1942 that the development of new weapons attracted any attention in the Reich. The advantage held by the Allied Nations was by that time already so great that there was no longer any hope of reversing it. The most important new weapon developed by the Germans during the war, the V2 long-range rocket, came into use at a stage of the conflict when Germany's position had become quite desperate.

The indifference of Hitler and those about him to research in natural science amounted to positive hostility.* It had at a very early period cost the Führer the good will of physicists. Only a handful of them, out of ambition or because they had failed to make their mark prior to the advent of the Third Reich, offered Hitler their full co-operation.† The great majority, however,

* The only exception to the lack of interest shown by authority was constituted by the Air Ministry. The Air Force research workers were in a peculiar position. They produced interesting new types of aircraft such as the Delta (triangular) and "flying discs." The first of these "flying saucers," as they were later called—circular in shape, with a diameter of some 45 yards—were built by the specialists Schriever, Habermohl and Miethe. They were first airborne on February 14, 1945, over Prague and reached in three minutes a height of nearly eight miles. They had a flying speed of 1250 m.p.h. which was doubled in subsequent tests. It is believed that after the war Habermohl fell into the hands of the Russians. Miethe developed at a later date similar "flying saucers" at A. V. Roe and Company for the United States.

† These were mainly engineers. Those engaged on theoretical research only co-operated in exceptional cases. It is true that the Allied Nations technicians, after the occupation of Germany, found thousands of new

were soon whispering to one another the inversion of a current Nazi slogan: "War must be harnessed to the service of science." Hitler's attempt to raise Germany to the status of a global power had been too lightly undertaken to succeed. Such was the view of men who never expected an experiment to come to anything unless its foundations had been well and truly laid. It consequently became important to save from the imminent disasters ahead as much as possible of German research that had not yet been utterly ruined by the regime. After the war had been lost science would probably be one of the few items still standing to the credit of Germany on her balance sheet.

Four factors combined to frustrate the construction of a German atom bomb. In the first place the absence of the eminent physicists driven into exile by Hitler proved a severe handicap. Second, the poor organization by the Nazis of research in the interests of war and its inadequate recognition by their government. Third, the insufficiencies of the technical equipment at hand for so complex a project. Finally (and too often overlooked so far), the actual personal attitudes of the German experts in atomic research counted against success. In the face of incomprehension by the authorities, they did nothing to overcome this obstacle. They did not push for the construction of such a bomb (in clear contrast to the German rocketeers, who finally conquered Hitler's indifference to "guided missiles" and got "their" V2 weapons). On the contrary, these physicists were able successfully to divert the minds of the National Socialist Service Departments from the idea of so inhuman a weapon.

Very little about this personal attitude has hitherto been revealed to the public. Most of those concerned have themselves preferred, for the sake of discretion and tact, to restrict mention of this delicate affair to a somewhat narrow circle. They content themselves with emphasizing, when called upon to explain why no German atom bomb existed at the end of the war, the lack of interest among their political leaders and the techni-

military inventions there. But these always concerned the practical application and technical improvement of scientific principles already well known.

cal difficulties which in fact became almost insuperable from the end of the year 1942 with the start of heavy Allied air attack. Heisenberg, who was in charge of the German uranium project, stated at the end of 1946, in the periodical *Naturwissenschaften*, that "external circumstances" had relieved the German experts in atomic research from the need "to take the difficult decision whether or not to produce atom bombs." This is a fair statement, but it is really only valid from the summer of 1942 on. But what happened before then? What did Heisenberg mean when he wrote in the same article: "German physicists worked consciously from the beginning to retain control of the project in view. They used their influence as experts on the subject to direct studies of it in the sense described in the present report."

At first the German "U Project," as it was called by the authorities, made more rapid progress, in the purely administrative sense, than similar efforts in the Allied countries and the still-neutral United States. Although most of the physicists were called up for military service immediately after war had broken out, three or four weeks later the more important of them were returned to their institutes as "indispensable." As early as September 26, 1939, more than a fortnight before Alexander Sachs could obtain an interview with Roosevelt and show him Einstein's letter, nine nuclear physicists—Bagge, Basche, Bothe, Diebner, Flügge, Geiger, Harteck, Hoffmann and Mattauch—attended a meeting at the Heereswaffenamt (the Army Weapons Department) in Berlin. At this conference a detailed program of work was drawn up and, as Diebner recollects, "separate tasks were assigned to various study groups." Such was the true origin of the so-called "Uranium Verein" ("Uranium Society"). Four weeks later a larger group met, by that time including Heisenberg and Weizsäcker. One of the first questions to be decided was the degree of refinement of the uranium oxide required for experimental purposes. The specialist appointed to carry out the chemical tests at Göttingen was, however, on military service. Some time passed before he could again be made available. Then it turned out that nearly all the uranium oxide in Germany had already been bought up by another Army department, which refused to return it under any circumstances. The department, in fact, was hoping to be able to produce armor-piercing shells by the use of this heavy metal as an alloy.

The first practical experiments, carried out at Leipzig, failed. The physicist Döpel, in ignorance of the chemical properties of uranium, had applied a metal shovel to it and thereby caused spontaneous combustion. When he poured water on the flames they spread faster than ever. A fountain of glowing uranium shot up, twenty feet high, setting the ceiling of the laboratory on fire. The Leipzig Fire Brigade turned out in full strength, and as a result only a few people were slightly burned. Döpel uttered a prophecy which at the time sounded merely melodramatic: "Hundreds more will fall for the supreme goal—the atom bomb."

In the autumn of 1939 the Kaiser Wilhelm Institute for Physics was made the scientific center of the Uranium Society. Peter Debye, the director of the Institute, was a Dutchman who had been working unmolested in Germany ever since 1909. He was now required either to become a German citizen or at least to publish a book in favor of National Socialism to prove his reliability. He rejected these insolent demands with contempt and took advantage of an invitation to give some lectures in the United States to turn his back for ever on his second home. Shortly afterwards Heisenberg succeeded him in charge of the Institute, a post he held for the duration of the war. This decision on Heisenberg's part was much criticized by his physicist friends abroad. It seemed to them a startling confirmation of the suspicion they had entertained for some time that Heisenberg had made his peace with Hitler.* Even in Germany itself

* Von Weizsäcker remarks in this connection: "After Debye's departure we came under the control of the Army Weapons Department. We had more and more extremely uncongenial people planted on us as time went on . . . we summoned Heisenberg to the Institute every week to advise us. After the lapse of a year he became, as we had foreseen, in practice the director of all our studies. We then succeeded in persuading the President and Senate of the Kaiser Wilhelm Society, who knew per-

Heisenberg's conduct was bitterly resented by certain physicists. They believed then and still believe today that if he had kept unequivocally clear of National Socialism he would not only have encouraged all those scientists of ability who were opposed to Hitler but might also have inspired them, as a leading spirit, to undertake active resistance.*

Heisenberg's friend and colleague Weizsäcker seeks to excuse him on the ground that he had always been a man of cosmopolitan training and outlook who nevertheless loved his native land. He is alleged to have remained in Germany in order to contribute to the survival of German physics under the calamities he foresaw.

But there was a further motive involved, perhaps the most important of all. Heisenberg only hints at it in his 1946 article. He and his closest friends wished, by controlling the Kaiser Wilhelm Institute for Physics, to keep the development of German atomic research in their hands. For they still feared at that time that other less scrupulous physicists might in different circumstances make the attempt to construct atom bombs for Hitler. It was considered certain not only in New York but also in Dahlem that the possession of such a weapon by a fanatical

---

fectly well what our political opinions were, to appoint Heisenberg our official director. This put an end to the nuisance of intruders from outside. In order to avoid encroaching upon Debye's prerogative, Heisenberg was given the title of 'Director *at* the Institute.' For we continued to consider Debye Director *of* the Institute."

* Heisenberg's own defense of his behavior at that time was expressed to the author as follows: "Under a dictatorship active resistance can only be practiced by those who pretend to collaborate with the regime. Anyone speaking out openly against the system thereby indubitably deprives himself of any chance of active resistance. For if he only utters his criticism from time to time in a politically harmless way, his political influence can easily be blocked. . . . If, on the other hand, he really tries to start a political movement, among students for instance, he will naturally finish up a few days later in a concentration camp. Even if he is put to death his martyrdom will in practice never be known, since it will be forbidden to mention his name. . . . I have always . . . been very much ashamed when I think of the people, some of them friends of my own, who sacrificed their lives on July 20 and thereby put up a really serious resistance to the regime. But even their example shows that effective resistance can only come from those who pretend to collaborate."

dictator who would stop at nothing would bring unimaginable misery upon the world.

By the winter of 1939–40 Heisenberg had already completed theoretical studies laying down the difference in principle between a uranium pile, in which the chain reaction is controlled, and a uranium bomb, in which the avalanche of neutrons is allowed to accelerate to the point of explosion. On July 17, 1940, Weizsäcker committed certain ideas to paper, under the heading "A possible method for the production of power from U 238." They showed that an entirely new substance, which could be used "as an explosive," might arise in a uranium pile. But he did not at that time call this substance plutonium, as did his colleagues in the Anglo-Saxon countries. He simply called it "Element 93," though he still retained a doubt whether it ought not really to be "Element 94."

Such ideas did not, however, go beyond the most intimate circle of Heisenberg's collaborators; they prudently refrained from passing on their preparatory theoretical studies on this theme. They were determined not to attract the attention of even their closest associates to the possibility of an atom bomb. Whenever suggestions pointing in that direction did nevertheless occasionally emanate from other physicists, they were not indeed rejected by Heisenberg as impossible in principle, but merely labeled unrealistic: "At present we can see no practicable technical method of producing an atom bomb during the war with the resources available in Germany. But the subject, nevertheless, must be thoroughly investigated in order to make sure that the Americans will not be able to develop atom bombs either." Such was the reason given, by this extremely influential group within the Uranium Society, for their merely expectant attitude. On the other hand, it was still considered necessary for the uranium project to retain, in the eyes of the government, enough of its highly promising character to justify the release of young physicists from military service. So this somewhat dangerous game, a source of mistrust and misunderstanding, continued, with its alternate postponements and promises.

In addition to Heisenberg and Weizsäcker, a third physicist, working in Germany during the years 1940 and 1941, had discovered that a uranium bomb could probably be manufactured very soon after a previous production in a uranium pile of a new explosive element. It was the man who had participated in the discovery of thermonuclear processes in the sun, Fritz Houtermans.

On Hitler's accession to power he had emigrated and thereafter been caught in the vortex of espionage psychosis that affected Soviet Russia in 1937. But he had been clever enough to postpone his "liquidation" by the Russian secret police through an adroit move on the chessboard of intrigue. He soon found that his examiners were not convinced by his protestations of innocence. In fact, they took no account of them whatever and simply went on torturing him. On one occasion he was beaten up for seventy-two hours on end, having all his teeth knocked out in the process. He accordingly contrived a trick. He told himself: "The officials in charge of the investigation are only concerned to produce their quota of such and such a number of confessions. I'll let them have what they want. But I shall insert a little time bomb in my statement. It may perhaps actually secure my release from prison." He proceeded to "confess" that he had in fact, as the secret police suspected, practiced espionage and sabotage against the Soviet Union, using for the purpose a certain small apparatus which he had himself invented in secret. Houtermans not only gave an accurate description of the instrument in question but also prepared detailed blueprints of its design. He alleged that it had enabled him to identify precisely, from the ground, the speed of Soviet aircraft, and thus to pass on important armament secrets to the Germans. The captive physicist was gambling on the chance that the plans he submitted might be laid before his former colleague Kapitza for investigation. The latter, or indeed any other expert, would then immediately perceive that Houtermans's "invention," in the form given, was perfectly senseless scientifically, and that consequently his entire confession had been untrue and must in all probability have been extorted from him by ill-treatment. In this way Houtermans not only gained a temporary respite from the attentions of his examiners but also a chance that his case might possibly be accorded fresh consideration by external authority.

In the spring of 1940 Houtermans was conditionally released. It is not known whether he owed his freedom to the petitions of well-known foreign physicists or to intervention by Kapitza. He was handed over, at Brest-Litovsk, to the Gestapo, who at first locked him up again. But eventually, on the intercession of von Laue, he was paroled subject to the proviso that he remain under observation by the Gestapo during the war and did not engage in any state or university research project. Houtermans had only been at liberty for a few days in Berlin when he learned the strictly kept secret of the existence of the Uranium Society. He instantly guessed what object the group in question probably had in mind, though owing to the years he had spent in prison he had not been able to keep in touch with the latest developments of atomic research. He himself had called attention, as early as 1932, to the possibility of a chain reaction and had worked, up to the time of his arrest in the Soviet Union, at the further elucidation of such problems. In 1937 he had actually lectured on neutron absorption to the Soviet Academy of Sciences. If the Communist secret police had not just then carried him off in the midst of his studies, it is quite possible that atomic fission and chain reaction would have been first discovered in Soviet Russia.

On hearing that Heisenberg and Weizsäcker were now seriously investigating the question of a practical application of the chain reaction Houtermans was profoundly shocked. He consulted von Laue. The Nobel prize winner consoled him with the remark: "My dear colleague, no one ever invents anything he doesn't really want to invent."

Since Houtermans was prohibited from working in any state-controlled or university laboratory, he applied, in May 1940, to the well-known inventor Baron Manfred von Ardenne, who was carrying out studies on behalf of the Post Office in his private research institute at Lichterfelde near Berlin. It is characteristic of the rivalry at that time prevailing among the various government offices of the Third Reich that in addition to the Ministry of Education and the War Office even the Post Office was engaged on atomic research. Ohnesorge, the Postmaster General, probably believed that he would rise enormously in his Führer's estimation if he of all people, though in charge of a civilian department, were to present Hitler, one day, with a "miracle weapon." But when in 1944 the long-yearned-for moment at last arrived and Ohnesorge, at a meeting of the Cabinet, started to hold forth about the present position of his studies in connection with the provision of a uranium bomb, Hitler interrupted him with the contemptuous observation: "Look here, gentlemen! You are all racking your brains to discover how we are going to win the war and, lo and behold, here comes our Postmaster, of all people, with a solution of the problem!" So far as the Führer was concerned, that settled the matter.

Houtermans did not dare to refuse point-blank the order of his new chief, von Ardenne, to apply himself to the uranium problem. Averse as he was to undertaking any sort of war work, his bitter experiences in the Soviet Union had taught him that in such cases the safest thing to do was to "play ball," to make a pretense of co-operation. He had to take care to keep memoranda composed in the course of the work locked away in secret safes.

In September 1940 Houtermans completed his first study of the uranium problem. His notes already at that time referred to the use of uranium piles to generate microscopically small quantities of Element 93 or 94. By July 1941 it was perfectly clear to him that it would be possible to produce an atom bomb provided that ponderable amounts of that substance—later called plutonium—could be generated in the uranium pile. Houtermans, however, did not report on this aspect of his work, since he did not wish to call the attention of the state authorities to the possibility of the construction of atom bombs. Moreover, he requested Dr. Otterbein, through whom the Post Office kept in touch with the Uranium Society, to see to it that for the present there should be no publication of his studies in the secret reports of the Army Weapons Department. By making further inquiries from time to time he was able to make sure that his notes continued to repose in a post-office safe. It was not until he learned in 1944 that the Hamburg physicist Harteck, independently of himself, had suggested the same possibility, that Houtermans consented to restricted publication of his article.* At this period of daily air raids on the country there was in any case no longer any question of the German project succeeding. Such is the explanation of the fact that at the end of the war one of the most important publications relating to German nuclear research was found among the secret "Research Reports of the Post Office." It was entitled: "Problem of the Release of Nuclear Chain Reactions."

Houtermans, despite his own political peril, naturally got into direct communication with Heisenberg and Weizsäcker, soon after his release by the Gestapo. He wanted to learn from their own lips just what the object of the Uranium Society was. He received the consoling information that every effort was to be concentrated on the problem of a "uranium machine" with the intention of progressively diverting the attention of the government departments from the possibility of the bomb. In the winter of 1941 he had a further confidential interview with Weizsäcker. He informed him of his own studies with Ardenne and said that he had kept quiet about the possibilities which those studies had established of the construction of an atomic weapon. Houtermans's confession encouraged his companion himself to be more frank than before. After a long discussion the two men agreed that the first and most important task of "uranium policy" must be to keep the departments in the dark about the now imminent feasibility of manufacturing such bombs. Heisenberg and Weizsäcker also assured Houtermans that they would treat his own studies accordingly if they ever came across them officially.

* Harteck had been a pupil of Rutherford. Shortly before the end of the war the Hamburg experimental physicist P. Koch denounced him to the Gestapo for "sabotage of atomic research." Koch committed suicide after the Allies entered Germany.

In addition to these three men there were at that time at least ten other prominent German physicists who had agreed that they must try to avoid working with Hitler's war machine or to make only a pretense of doing so. The names of German physicists unwilling to supply Hitler with supplementary armaments were deposited, after the war had begun, in Sweden—with Professor Westgren—and in Holland—with Professor Burgers. It was considered that an open strike of research workers would be dangerous, as it would leave the field open for unscrupulous and ambitious persons. So long as a policy of delay and postponement proved practicable, it was resolved that the risk should be taken. But at least some atomic scientists considered that when such a course could no longer be followed it would be their duty to take an active part in politics. They therefore kept in touch with the party of conspirators led by General Beck and the Chief Burgomaster of Leipzig, Gördeler.

These physicists, some of whom had decided upon a policy of passive resistance only after an extremely hard struggle with their consciences, did not form by any means an organized and compact body. They just knew who "belonged." If a newcomer attached himself to the circle, he was discreetly sounded. The process began with the exchange of harmless political jokes, continued with certain criticisms, at first mild, of the regime and gradually approached more and more dangerous topics. The well-known nuclear physicist Haxel remembers: "By slow degrees more and more pledges of mutual confidence were given on both sides till eventually each of us, so to speak, carried the other's life in his hands. At that moment we began at last to talk freely together."

The favored meeting spot for the heretical German atomic scientists was either the Berlin office of Dr. Paul Rosbaud, the publisher of scientific books and periodicals, or his modest home in the suburb of Teltow. Rosbaud, a temperamental Austrian who knew most of his authors intimately, showed a courage bordering upon recklessness in dealing with the Gestapo. If anyone may claim to have been the soul of passive resistance of German scientists to Hitler, it is this warm-hearted man. In the midst of war he upheld not only with words but also with deeds the idea of the solidarity of all men of good will. He had a habit, for example, of entering, as often as he could manage it —"by mistake"—the compartments on the metropolitan railway reserved for foreigners on forced labor. There he would secretly distribute provisions or other small gifts. In the middle of the war he arranged for the French physicists Pérou and Piattier to be allowed to translate into French, for the firm of Julius Springer, a famous German book on physics. Permission was obtained for them to be released from their prison camp to undertake the task. But Rosbaud had previously also obtained Joliot-Curie's consent, to prevent the French officers from being later charged with collaboration on account of their work as translators.

Many channels of communication remained open, even during the war, between Joliot and the German physicists hostile to the Nazis. In the summer of 1940 Wolfgang Gentner, who had worked with Joliot-Curie in peacetime, took over his former chief's institute, on behalf of the German military authorities, in occupied Paris. But he only did so after Joliot had expressly signified his assent to this proceeding. The two atomic scientists, whose trust in each other had not been in the least diminished by the war, then sat down, as they often had in former days, on the terrace of a café in the boulevard Saint Michel. There they drew up, on the back of a menu, the draft of a solemn agreement to the effect that Joliot's laboratory was not to be used in any circumstances for research in the interests of the war. On many subsequent occasions Gentner found himself obliged to intervene personally to save Joliot—and also Paul Langevin—from the clutches of the SS. At last, in 1943, he was recalled from Paris on account of the "weakness" he had shown. He was replaced by a Nazi firebrand, who later had to be hunted for by the police in connection with diamond smuggling.

On a number of occasions the question arose among the passivist German atomic scientists whether, and, if so, how, information should be conveyed to the other side as to the condition of

German research work and the real intentions of the Uranium Society. Houtermans, who had suffered in the dungeons of the Stalinist secret police, entertained no doubts whatever on the matter. He argued that "every decent man confronted by a totalitarian regime ought to have the pluck to commit high treason." Heisenberg was probably incapable of taking up so radical an attitude. He was one of those people, according to Weizsäcker, "who had been so deeply shocked by the terror and cynicism with which the Nazis had started the war, that while, on the one hand, they could not bring themselves to desire a German victory, on the other hand they could wish as little for a German defeat, with all its frightful consequences."

Although Heisenberg may not have longed for the eventual German collapse, he was convinced, purely as a matter of logic, that Germany must lose. As he put it later in conversation: "For Germany the war was like an end game in chess in which she possessed one castle less than her adversary. The loss of the war was as certain as the loss of an end game under these conditions."

What could be done at this stage to lighten the shock of collapse and make the closing phase of the war less dreadful for Germany? Such, in all probability, was the question Heisenberg put to his conscience at the time when he made up his mind to discuss the atom bomb with an influential foreign friend. It might be possible, by passing on the negative secret that the Germans did not intend to construct an atom bomb, to stop the British or Americans doing so—if they had any intention to use such a weapon against Germany—and thus save the country the horrors of atomic bombardment. Such considerations may have prompted, at the end of October 1941, a little-known peace feeler. By the expedient of a silent agreement between German and Allied atomic experts, the production of a morally objectionable weapon was to be prevented.

Heisenberg had at that period been invited to give a lecture in occupied Copenhagen. He naturally looked up his old teacher and friend Niels Bohr on that occasion. Although Bohr, being half-Jewish, was in personal danger, he had remained in the Danish capital. He realized that his presence was the only protection that could be given to the "non-Aryan" members of his Institute. He had repeatedly been urged by Allied agents to escape, but he had told them emphatically that he would stay in Copenhagen as long as it proved at all practicable. Bohr's correspondence with foreign countries was even more eagerly scanned by his friends than by the Nazi censors. In a telegram which he sent to Frisch in England shortly after the occupation of Copenhagen Bohr had asked after "Miss Maud Rey at Kent." The recipient could not remember ever having even heard of such a lady. He accordingly hit upon the idea that the words probably represented an anagram for "radium taken." He drew the reasonable conclusion that Bohr had wished to convey surreptitiously the information that the Germans had confiscated the stocks of radium in his Institute. As a matter of fact it was found very much later that Bohr had meant precisely what he had written in the telegram. There was no secret message. He had really only been asking after an old friend. But the lady's name had been somewhat garbled in dispatch. The British physicists, however, who had been busy with their own atom project ever since 1940, knew nothing about that mistake at the time. They determined, on the receipt of the message, to give their plan, in honor of their master, the code-name M.A.U.D.

Shortly afterwards Bohr asked, on a postcard, for news of an old pupil of his named "D. Burns." On that occasion, too, a secret message of profound importance was conjectured to be concealed beneath the words. It was supposed that Bohr wished to inform his Allied colleagues that his studies had led him to discover that "D" (deuterium or heavy hydrogen) is inflammable.

The German physicists knew that Bohr, if he cared to do so, could at once establish the best possible relations between themselves and the atomic scientists who were working in Britain and the United States. For this reason he would be the ideal mediator.

But unluckily the important interview in Copenhagen between Heisenberg and Bohr was ill-starred from the beginning. It had been reported to Bohr that Heisenberg had defended, at a reception given in his honor shortly before, the German invasion of Poland. The fact was that Heisenberg, in order to disguise his true sentiments, was in the habit of expressing himself quite differently in society, especially abroad, from the way in which he talked in private. But Bohr, that fanatical devotee of truth, neither could nor wished to recognize such a double game, learned in the hard school of totalitarian compulsion. Accordingly, when Heisenberg came to see him, he at once assumed an extremely reserved and even chilly attitude towards the pupil who had once been his favorite.

Heisenberg began by entreating Bohr to realize the coercive pressure which had been brought to bear on the German physicists. Then he gradually and cautiously steered the conversation towards the question of the atom bomb. But unfortunately he never reached the stage of declaring frankly that he and his group would do everything in their power to impede the construction of such a weapon if the other side would consent to do likewise. The excessive prudence with which both men approached the subject caused them in the end to miss it altogether. When Heisenberg asked whether Bohr considered it probable that such a bomb could be constructed, the latter, who had heard nothing more since April 1940 about the progress—which had of course been kept secret—of atomic research in Britain and America, replied in guardedly negative terms. Thereupon Heisenberg nerved himself to assure the other with all the emphasis at his command that he knew it to be perfectly possible to produce such a weapon and that it could actually be manufactured, if a very great effort were made within a very short time.*

Bohr himself was much troubled by Heisenberg's account of the position, so much so that he paid less than the right amount of attention to Heisenberg's remarks later about the dubious moral aspect of such a weapon.

When Heisenberg took leave of his master he had the im-

* The first written reference by Heisenberg to this interview was made in a letter to the author which is printed in the note at the end of this chapter.

pression—which later events were to prove correct—that the conversation had made matters worse rather than better. Bohr's mistrust of the physicists who had remained in Hitler's Germany had not been lessened by his pupil's visit. On the contrary, he was now convinced that the men in question were concentrating intensively and successfully on the manufacture of a uranium bomb.

To correct this false impression, a second German atomic physicist traveled to Copenhagen, shortly afterwards, to see Bohr. Meanwhile the latter's suspicions had deepened to such an extent that when young Jensen frankly stated what Heisenberg, in his excessive prudence, had only hinted at, Bohr merely believed his visitor to have been sent as an *agent provocateur*.

As soon as the occupation of Bohr's Institute became imminent, he escaped, in 1943, to England via Sweden, making the journey under dangerous conditions. On his arrival in England, as a result of these interviews, he proceeded to encourage the Anglo-American authorities in their efforts to anticipate Hitler in the construction of an atom bomb.

Note

In a letter to the author Heisenberg gave this version of his conversation with Bohr:

As far as I remember, although I may be wrong after such a long time, the conversation came about as follows. My visit to Copenhagen took place in autumn 1941, actually I think it was at the end of October. At this time we in the "Uranverein," as a result of our experiments with uranium and heavy water, had come to the following conclusion: "It will definitely be possible to build a reactor from uranium and heavy water which will provide energy. In this reactor (in accordance with theoretical work of von Weizsäcker) a decay product of uranium 239 will be produced which will be as suitable as uranium 235 as an explosive in atomic bombs." We did not know a process for obtaining uranium 235 with the resources available under wartime conditions in Germany, in quantities worth mentioning.

On the other hand, as the production of the atomic explosive can only be accomplished in huge reactors which have been working for years, we were convinced that the manufacture of atomic bombs was possible only with enormous technical resources. We knew that one could produce atom bombs but overestimated the necessary technical expenditure at the time. This situation seemed to us to be a favorable one, as it enabled the physicists to influence further developments. If it were impossible to produce atomic bombs this problem would not have arisen, but if they were easily produced the physicists would have been unable to prevent their manufacture. This situation gave the physicists at that time decisive influence on further developments, since they could argue with the government that atomic bombs would probably not be available during the course of the war. On the other hand there might be a possibility of carrying out this project if enormous efforts were made. Further developments proved that both attitudes were to the point and fully justified—as the Americans for instance actually could not employ the atomic bomb against Germany.

Under these circumstances we thought a talk with Bohr would be of value. This talk took place on an evening walk in a district near Ny-Carlsberg. Being aware that Bohr was under the surveillance of the German political authorities and that his assertions about me would probably be reported to Germany, I tried to conduct this talk in such a way as to preclude putting my life into immediate danger. This talk probably started with my question as to whether or not it was right for physicists to devote themselves in wartime to the uranium problem—as there was the possibility that progress in this sphere could lead to grave consequences in the technique of war. Bohr understood the meaning of this question immediately, as I realized from his slightly frightened reaction. He replied as far as I can remember with a counterquestion. "Do you really think that uranium fission could be utilized for the construction of weapons?" I may have replied: "I know that this is in principle possible, but it would require a terrific technical effort, which, one can only hope, cannot be realized in this war." Bohr was shocked by my reply, obviously assuming that I had intended to convey to him that Germany had made great progress in the direction of manufacturing atomic weapons. Although I tried subsequently to correct this false impression I probably did not succeed in winning Bohr's complete trust, especially as I only dared to speak guardedly (which was definitely a mistake on my part), being afraid that some phrase or other could later be held against me. I was very unhappy about the result of this conversation.

Heisenberg, however, says that he is unable today to recollect the precise wording of this talk. If one could interpret the content of this conversation in psychological terms, it would depend on very fine nuances indeed. As neither of the two participants possesses a stenographic record, Heisenberg's notes, done from memory—with all their reservations—are the best existing source.

Seven

The Laboratory Becomes a Barrack (1942-1945)

It seems paradoxical that the German nuclear physicists, living under a saber-rattling dictatorship, obeyed the voice of conscience and attempted to prevent the construction of atom bombs, while their professional colleagues in the democracies, who had no coercion to fear, with very few exceptions concentrated their whole energies on production of the new weapon.

Fifteen years later a German scientist tried to explain this situation in the following words: "We were really no better, morally or intellectually, than our foreign colleagues. But by the time the war began we had already learned from the bitter experience of nearly seven years under Hitler that one had to treat the state and its executive organizations with suspicion and reserve. The citizens of totalitarian countries are rarely good patriots. But our colleagues elsewhere had at that time complete confidence in the decency and sense of justice of their governments." The speaker hesitated for a moment and then added: "I doubt, incidentally, whether exactly the same situation prevails in those countries today."

On the outbreak of war there was a general movement among scientists in the nations threatened by Hitler to support their governments. This was a magnificent vote of confidence in the justice and moral responsibility in practice of the democratic system. It was all the more astonishing because the scientist is not at heart an ideal citizen but one never satisfied, eternally striving for what is new. It is his nature to call in question the existing order of things and seek for new and improved solutions of problems. This tendency of scientists to criticize and amend could be much better understood by a democratic government than by a rigidly authoritarian and totalitarian regime. In fact it was of positive advantage to a democracy. It was invariably true at that time that the conservative element was chiefly represented by the military authorities. Their opposition to all new weapons was great, but never so strong as in the atom project. American nuclear physicists were still chuckling, years later, over stories illustrative of the widespread mistrust and shortsightedness at first evident among the representatives of the armed forces about the plans of "those fools." Senior officers who had grown gray in the service of their country gave those whom they interviewed and to whom they were bound to listen in obedience to commands from above clearly to understand that they had little more respect for their visitors than they had for other crazy inventors who were smothering the Army and the Navy with designs for every conceivable type of wonder-working weapon, most of them utterly impracticable. "Only the other day," an American officer growled at a group of atomic scientists during one such conference, "someone sent me another of those death-ray generators. I tried the thing out on our regimental mascot, a he-goat. The brute's still as lively as ever!"

Nevertheless, the nuclear physicists gradually succeeded, with the help of civilian officials and politicians, in overcoming the resistance of the Colonel Blimps. Their success proved most rapid in France. After the outbreak of war, when Joliot-Curie called on the Minister of Munitions, Raoul Dautry, and told him about the possibility of atomic weapons, Dautry, a former industrialist who had always previously kept his eyes open for the acquisition of new technical devices, positively reproached his visitor for not having come to see him before. Joliot's account of the significance of uranium and heavy water in atomic research caused Dautry to act at once. At the time of the German invasion France not only possessed larger stocks of uranium oxide than any other country in the world, but also all the heavy water existing in Europe—only 185 kilograms (3.6 cwt.) in all, which had been bought from the Norwegian firm Norsk Hydro in March 1940 by the French special commissioner Jacques Allier and brought to France by air in twelve sealed aluminum containers.

On May 16, 1940, the telephone rang in the office of Henri Moureux, Deputy Director of the Nuclear Chemistry Laboratory at the Collège de France. Joliot-Curie was on the line. He requested his collaborator to come and see him immediately. On his arrival he announced with visible excitement: "The front has been penetrated near Sedan. Dautry has just telephoned to me. The heavy water will have to be taken to a safe place at once." That same night the containers of the valuable "Product Z," as it was known in code, were sent to central France. They were stored in the underground depositories of a branch of the Bank of France at Clermont-Ferrand.

On June 10, 1940, Joliot and his closest associates had already begun burning all the papers which gave any indication of the state of his atomic research, to leave the Germans, already quite near Paris, no information about his work. It was a precautionary measure which unfortunately proved useless. A few days later copies of these documents, together with many other files of the French War Office, fell into the hands of German troops at Charité-sur-Loire.

Joliot stayed in Paris. He did not want to be separated from his precious laboratory apparatus, which included a recently completed cyclotron, the first ever made in Central and Western Europe. He entrusted to his collaborators Halban and Kowarski, who had set up a temporary refuge for the nuclear physics and nuclear chemistry center in the villa Clair Logis at Clermont-Ferrand, the task of transferring the heavy water to England via Bordeaux. Halban tells the story:

"One night we deposited our valuable consignment for safe keeping in the State Prison at Riom. The place in the prison which was most secure from intrusion was the death cell, which was cleared for the time being to make room for our water bottles. The dislodged convicts under sentence of death themselves carried the heavy containers into the cell. Next morning the Governor of the prison, probably already in fear of the new masters, refused to release the deposited articles. Dautry's special commissioner had to threaten him with a drawn revolver before he would give them up. Then we were able to continue our journey."

After many hazardous incidents the heavy water reached Bordeaux and was put aboard the British collier *Broompark*. The fugitive atomic experts were aided by the Earl of Suffolk, Scientific Attaché at the British Embassy in Paris. This adventurous English peer had once, in his youth, run away from home and signed on aboard ship. He had learned carpentry in those days and it now stood him in good stead. In great haste he built a seaworthy raft aboard which the aluminum canisters containing the precious Product Z were securely stowed, together with industrial diamonds valued at 2½ million pounds sterling. Halban and Kowarski solemnly promised the Earl that if the *Broompark* came to grief for any reason whatever, by striking a mine or being hit by air bombardment, they would get the raft out of the hold, launch it and, come what might, stick to it in the open sea. But this precaution eventually proved to have been unnecessary. The ship, with its strategically important cargo, reached England in safety. A vessel which left Bordeaux at the same time as the collier was sunk, and Joliot tricked the German counterespionage into believing that the heavy water had been aboard the lost ship.

Work on the problem of atomic fission had begun also with government support in England. No sooner had George P. Thomson, Professor of Physics at the Imperial College in London, read in *Nature*, in the spring of 1939, the studies by Joliot and his associates on the phenomena of neutron emission, than he got into touch with Henry (now Sir Henry) Tizard, who had been in charge of the Research Department of the Royal Air Force since 1934. With Tizard's recommendation Thomson called at the Air Ministry. He reported later that he had felt like a character in a third-rate thriller while he was explaining to the officials at the Ministry what tremendous consequences these discoveries might have for the development of power in industry and war. His sensational revelations were however taken quite seriously and he was offered, though in somewhat skeptical terms, a ton of uranium oxide and a small sum of money to assist him in his researches. This ready response may have been influenced by certain information which had been received concerning the meeting of German atomic experts at the Reich Ministry of Education. The news had been brought from Berlin by the English physicist R. S. Hutton in the middle of May. It is true that Thomson was informed, on the outbreak of war, that he could expect no priority of supplies for his experiments. There were more urgent matters to be dealt with just then.

It was mainly foreign physicists not cleared by security who were consulted about the work on atomic fission, regarded as "not essential to the war effort," which was being directed by British subjects, Thomson, Chadwick and Feather. The first of these foreigners were Frisch, who had just escaped from Copenhagen to England, Rudolf Peierls, Joseph Rotblat, Franz Simon and eventually, after their flight from France, Halban and Kowarski. The best-known refugee physicist, Max Born, was teaching in Edinburgh. His wife, a Quaker, had persuaded him to have nothing to do with any war work. Shortly afterwards one of his most talented students joined Peierls's team. This was Klaus Fuchs, a German parson's son who had fled to England via Paris. He took a leading part in the calculations of the critical size of the bomb.

In the United States only very slow progress was at first made in the field of atomic research. It was nearly ten weeks before Alexander Sachs at last found an opportunity, on October 11, 1939, to hand President Roosevelt, in person, the letter composed by Szilard and signed by Einstein at the beginning of August. In order to ensure that the President should thoroughly appreciate the contents of the document and not lay it aside with a heap of other papers awaiting attention, Sachs read to him, in addition to the message and an appended memorandum by Szilard, a further much more comprehensive statement by himself. The effect of these communications was by no means so overpowering as Sachs had expected. Roosevelt, wearied by the prolonged effort of listening to his visitor, made an attempt to disengage himself from the whole affair. He told the disappointed reader that he found it all very interesting but considered government intervention to be premature at this stage.

Sachs, however, was able, as he took his leave, to extort from the President the consolation of an invitation to breakfast the following morning. "That night I didn't sleep a wink," Sachs remembers. "I was staying at the Carlton Hotel. I paced restlessly to and fro in my room or tried to sleep sitting in a chair. There was a small park quite close to the hotel. Three or four times, I believe, between eleven in the evening and seven in the morning, I left the hotel, to the porter's amazement, and went across to the park. There I sat on a bench and meditated. What could I say to get the President on our side in this affair, which was already beginning to look practically hopeless? Quite suddenly, like an inspiration, the right idea came to me. I returned to the hotel, took a shower and shortly afterwards called once more at the White House."

Roosevelt was sitting alone at the breakfast table, in his wheel chair, when Sachs entered the room. The President inquired in an ironical tone:

"What bright idea have you got now? How much time would you like to explain it?"

Dr. Sachs says he replied that he would not take long.

"All I want to do is to tell you a story. During the Napoleonic wars a young American inventor came to the French Emperor and offered to build a fleet of steamships with the help of which Napoleon could, in spite of the uncertain weather, land in England. Ships without sails? This seemed to the great Corsican so impossible that he sent Fulton away. In the opinion of the English historian Lord Acton, this is an example of how England was saved by the shortsightedness of an adversary. Had Napoleon shown more imagination and humility at that time, the history of the nineteenth century would have taken a very different course."

After Sachs finished speaking the President remained silent for several minutes. Then he wrote something on a scrap of paper and handed it to the servant who had been waiting at table. The latter soon returned with a parcel which, at Roosevelt's order, he began slowly to unwrap. It contained a bottle of old French brandy of Napoleon's time, which the Roosevelt family had possessed for many years. The President, still maintaining a significant silence, told the man to fill two glasses. Then he raised his own, nodded to Sachs and drank to him.

Next he remarked: "Alex, what you are after is to see that the Nazis don't blow us up?"

"Precisely."

It was only then that Roosevelt called in his attaché, General "Pa" Watson, and addressed him—pointing to the documents Sachs had brought—in words which have since become famous:

"Pa, this requires action!"

The success of American atomic research during the war, which astonished the world, colored later descriptions of the subject. What was seen in retrospect as a difficult but straight road leading to its goal was really a labyrinth of winding streets and blind alleys.

Teller criticizes as follows one of these excessively rosy views of the early history of the American atom bomb: "There is no mention of the futile efforts of the scientists in 1939 to awaken the interest of the military authorities in the atomic bomb. The reader does not learn about the dismay of scientists faced with the necessity of planned research. He does not find out about the indignation of engineers asked to believe in the theory and on such an airy basis to construct a plant."

Wigner remembers the resistance. "We often felt as though we were swimming in syrup," he remarks. Boris Pregel, a radium expert, without whose disinterested loan of uranium the first experiments at Columbia University would have been impossible, comments: "It is a wonder that after so many blunders and mistakes anything was ever accomplished at all." Szilard still believes today that work on the uranium project was delayed for at least a year by the shortsightedness and sluggishness of the authorities. Even Roosevelt's manifest interest in the plan scarcely accelerated its execution. Sachs knew his way pretty well about the jungle of bureaucratic intrigue. At first he succeeded in preventing the Army or Navy from monopolizing the project. He proposed that Briggs, the director of the National Bureau of Standards, should be put in supreme control of the plan. But Briggs, though able enough, was a sick man, due at the time to undergo a serious operation. He could not take such energetic action as might often be necessary. It seemed for a while as though both he and the entire "Project S-I," as it was provisionally called, would take leave of their precarious lives simultaneously. But Briggs recovered and S-I, too.

Before the end of June 1940 there was no hope of obtaining any funds from the government for atomic research. On the contrary, criticism of a "plan that had no prospects of success" mounted in volume. A second letter from Einstein, dated March 7, 1940, called attention to the "intensification of German interest in uranium since the beginning of the war." But this communication did little good. It was not until the reports on the progress of British atomic research, which Dr. R. H. Fowler sent regularly to Briggs after July 1940, on the instructions of his government, became more and more positive that the official interest of Washington revived. By July 1941 a memorandum by the Thomson Committee, basing its opinion on the work being done in Britain, had declared that it seemed "quite probable that the atom bomb may be manufactured before the end of the war." And so at last, on December 6, 1941, just one day, as it happened, before the Japanese attack on Pearl Harbor and America's official entry into the war, the long-delayed decision was taken to apply substantial financial and technical resources, in earnest, to the construction of the weapon.

It was the scientists of European origin who devoted themselves with the greatest determination to the project. During its early stages they were severely handicapped by being "aliens" or even, like Fermi, who came from Italy, "enemy aliens." The trouble became so serious that Wigner, offended by the mistrust he encountered, informed Briggs in writing that in these circumstances he would unfortunately be obliged to withdraw from any work concerned with atomic-research development. His grievances were later satisfied and he became one of the most important collaborators in the plan. The British were much more generous in this respect. After some preliminary hesitation they treated the scientists who had taken refuge with them as entitled to the same rights as themselves. Weisskopf remembers that as a former Austrian he only obtained permission from the American authorities with difficulty to attend a special conference with three English gentlemen. The three visitors, when they arrived from London, turned out to be Halban, Peierls and Simon, who had all been, like Weisskopf himself, Central Europeans until a short time before.

The countless administrative and technical obstacles which blocked the road to the release of atomic energy were finally overcome simply and solely by the determination and obstinacy of the scientists resident in the Anglo-Saxon countries. They did much more than obey orders. They repeatedly took the initiative in bringing that mighty weapon into the world. Their initiative was perhaps the most important "raw material" used in the achievement of atomic power, but their enthusiasm, which surmounted every difficulty and was founded upon a passionate belief in the justice of the Allied cause, did not bring them much appreciation.

Many scientists were inspired at that time by the honest conviction that this was the best, in fact the only, way to prevent employment of the atomic weapon during the current war. "We must have some countermeasure available to meet any possible threat of atomic warfare by Germany. If we only had such a thing both Hitler and ourselves would be obliged to renounce the use of such a monstrosity." So said the few who were in on the secret.

The notion that the Germans already had a dangerous start in the atomic-armaments race had become so deep rooted that it was treated as a certainty. "We were told day in and day out that it was our duty to catch up with the Germans," recalls Leona Marshall, one of the few women in the project. There was never any doubt of this hypothesis. It silenced scruples that occasionally cropped up. In 1941 the chemical expert Professor Reiche, who had escaped from Germany a few weeks before, arrived at Princeton. He brought a message from Houtermans to the effect that the German physicists had hitherto not been working at the production of the bomb and would continue to try, for as long as possible, to divert the minds of the German military authorities from such a possibility. This news was passed on from Princeton to Washington by another scientist who had emigrated to America, the physicist Rudolf Ladenburg. But it does not seem ever to have reached those actually engaged on the atomic project. A year later Jomar Brun, former technical manager of the heavy-water works at Rjukan in Norway, who had fled to Sweden after the occupation by German troops in 1940, stated that he had been told by Hans Suess, the German atomic expert employed there, that production at Rjukan could not attain the dimensions important for war purposes in much less than five years. The Rjukan factory was, however, destroyed in a heroic but actually pointless raid carried out by Allied Commando troops and aircraft.

Was it possible that such reports were simply not believed, or did people not want to believe them? Otherwise, would not the zeal and speed of the work of those engaged on Project S-I have diminished?

In 1942 the Allied atomic project entered an entirely new phase. Roosevelt and Churchill agreed to concentrate the work of the British and American research teams in Canada and the United States. In the United States supreme control of atomic research was transferred from the scientists to a Military Policy Committee consisting of three members of the Armed Forces:

General Styer, Admiral Purnell and General Leslie Groves, and only two professional investigators, Dr. Vannevar Bush and Dr. James Conant. After August 13, 1942, the whole plan became known under the code name of either the DSM (development of substitute materials) or the "Manhattan Project." From then on the atomic experts were simply designated "scientific personnel" and obliged to submit to the strict rules of military secrecy.

It was probably the first time in history that so brilliant a group of minds had voluntarily undertaken to adopt a mode of work and existence so unlike their normal way of life. They accepted as obvious the rule that they were to publish no more of their discoveries until after the war. They had themselves, after all, been the first to propose, even before the war, that secrecy should be maintained. But the military authorities went much further than this prohibition. They erected invisible walls round every branch of research, so that no department ever knew what any other was doing. Barely a dozen of the total number of some 150,000 persons eventually employed on the Manhattan Project were allowed an over-all view of the plan as a whole. In fact only a very small number of the staff knew that they were working on the production of an atom bomb at all. For example, the majority of workers at the computing center in Los Alamos had for long no idea of the real purpose to be served by the complicated calculations they carried out on the computing machines. This ignorance prevented their taking any genuine interest in the job. At last Feynman, one of the young theoretical physicists, managed to obtain permission to tell these people what was going to be done at Los Alamos. Thereupon the services rendered by the department reached a considerably higher standard and some of the staff, from that time forward, worked overtime of their own accord.

In cases where the exchange of views between members of separate branches was found absolutely necessary special sanction for this step had to be obtained in the first place from the military directorate. The physicist Henry D. Smyth, subsequently author of the official report on the entire project, became involved, owing to these regulations, in a conflict of an unusual nature. Since he was in charge of two departments simultaneously, he would have been able, strictly speaking, to talk to himself only after having obtained prior permission to do so.

This so-called "compartmentalization" was decreed in spite of the fact that intensive security measures had already been taken by police investigation, cross-examination and questionnaires to check the previous private and political activities as well as the personal reliability of every collaborator in the plan, all of whose movements were subjected to a system of observation thought out to the smallest detail. Every inhabitant of the three "secret cities"—Oak Ridge, Hanford or Los Alamos—could only receive or dispatch correspondence through the censorship. If any passage in such a letter displeased one of the censors, he was not satisfied with the usual procedure of cutting out certain undesirable words. He simply returned the letter to the sender for redrafting, the object being to prevent the recipient at all costs from being mystified in any way. Telephone conversations were regularly listened to by a third party, and hotel porters in the neighborhood of each project were employed as counterespionage agents.

The most prominent of the atomic experts were provided with official bodyguards who followed them everywhere. In addition, a special watch was kept upon those who were not regarded, on political or any other grounds, as a hundred per cent reliable. Agents followed in their footsteps. Microphones were concealed in their offices and private residences to record their conversations. The head of the security division of the Manhattan Project, John Lansdale, Jr., admitted after the war that certain other methods of supervision and tricks were employed which have not been disclosed even today.

He considered them, however, so dishonorable that he only referred to them, when generalizing on the subject, as "nasty things." The scientists themselves were actually asked to render active assistance in the spinning of this web of prying. In addition to keeping silence about their work it was supposed to be their duty to tell lies to third parties about what they did and where they lived. Not even their closest relatives were to be given the merest hints as to where and in what business they were employed. Counterespionage turned every husband working in one of the laboratories into a Lohengrin who had to tell his Elsa, "You must never ask me any questions." Some of the scientists, naturally enough, would not allow the shadow of "security" to fall between them and their nearest and dearest. Wives who were in on the secret then had the difficult task of behaving, in the company of the wives of other physicists who submitted to the rule, as if they were equally unsuspecting.

The leading spirit in the management of the Manhattan Project was a professional soldier named Leslie Richard Groves. He was forty-six years old when he was entrusted, on September 17, 1942, with the administration of atomic research. There was one thing that worried "Greasy" Groves, as he had been known at West Point Military Academy. Throughout his professional career he had only held desk commands. For this reason he had remained a lieutenant for sixteen years and had only been appointed to the provisional rank of colonel since the outbreak of war. He had at last, on the eve of his nomination as head of the Manhattan Project, been offered an active service command, and was accordingly by no means delighted when his immediate superior sent for him and told him he had now been selected for a post at home, though it was the biggest job of the war and might be destined to make the decisive contribution to victory. General Groves—for this appointment bestowed upon him, by way of consolation, the rank he had been waiting for so long—had been considered for the position he was now to occupy because he had more experience of supervising building construction than any other officer in the Army. He had been in charge of the erection of a number of barracks. But in particular he had directed that of the gigantic new War Department Pentagon. He was now to conjure into being the secret "atom cities," with their laboratories, and administer them. Both externally and internally they assumed, under his direction, the aspect of barracks.

When Groves assembled his staff at Los Alamos for the first time, he began his speech—as was soon afterwards generally reported in the town—with the words: "Your job won't be easy. At great expense we have gathered here the largest collection of crackpots ever seen." Groves did not always trust the official tests and supervision imposed on the scientists. In one case he arbitrarily demanded of the War Department the instant internment, as an enemy alien, of a scientist of non-American origin who was working in Los Alamos, though the General admitted that he had not the least evidence against the man. When he was asked, accordingly, on what grounds he made his application, he answered that his "intuition" was good enough. He couldn't accuse the man of disloyalty or treason, Groves said, but just didn't trust him and considered him a "detriment to the project."

The Secretary of War adhered to the general rule in democratic countries that no one may be accused or condemned on insufficient evidence. He refused to sanction the "protective custody" proposed by Groves. Groves regarded this decision as simply one more proof of the credulousness and laxity of the civil authorities. He later claimed that whenever possible he acted on his own responsibility, against the will of Washington. In 1954 he boasted: "I was not responsible for our close co-operation with the British. I tried to make it as difficult as possible."

The enterprising General and those whose enterprise was confined to the atomic field were too diverse in character to be able to come to any genuine mutual understanding. Groves felt—as he still feels—that they underestimated his intelligence. Accordingly, he repeatedly tried to prove to them that his ability was at least equal to theirs even in their own sphere. "The first time we had any serious discussion," he relates, "in the newly established Metallurgical Laboratory at Chicago, I caught the gentlemen in a mistake in arithmetic. They sure didn't fool me. There were a few Nobel prizewinners among them. But I showed them, just the same, where they were wrong and they couldn't deny it. They never forgave me for that."

In reality "Gee Gee," as his irreverent recruits with high intelligence quotients called him, was by no means despised, but rather admired, by those very persons. It was not so much his mathematical gifts, of which he himself was so proud, that they appreciated, as his undeniable tenacity and stubbornness. The atomic scientist Philip Morrison states: "For a time I worked next door to one of his many offices and was amazed one day to hear him discuss the need for the acquisition of a tennis net with the same seriousness and for just as long as that for the expenditure of a million dollars on some experiment of quite uncertain prospects. In the end he refused to throw away the few dollars required to buy the tennis net but sanctioned the million to finance the experiment. I believe he would have fenced in the moon if he had been told it would facilitate our project."

It was not always easy for Groves to defend his decisions, which often seemed pointless. "Why build a road with eight traffic lanes to the plutonium works at Hanford?" asked the roadmen, who were miserably housed in adjoining barracks. "It's sheer waste of money!" But Groves could not tell them that his road, though admittedly expensive, was a security measure. Two or four lanes would have been enough for ordinary traffic. In the case of an explosion, always possible, the eight lanes would only just suffice to rescue the men at work and the families living near as quickly as possible from the radioactive fumes.

Groves also occasionally took a false step, as was inevitable in the performance of so great and novel a task. His critics are able to list quite a number of such erroneous decisions, taken too hurriedly or without considering all the factors involved, but the General is even today, in his disarming naïveté, convinced of the contrary. "You ask me why I haven't yet written my memoirs?" said he to a visitor eleven years after the war. "Well, it's simply because I was always right. And no one would ever believe that or even forgive me for it."

The scientists only pretended, of course, right from the start, to observe the compartmentalization which Groves was so particularly anxious to maintain. After the war Szilard gave the following account of the matter to a committee of Congress: "These kinds of rules could not be obeyed if you wanted to obey them. But we did not want to obey them because we had to choose between obeying these rules and sabotaging or slowing down our work, and we used common sense in place of obeying rules. Hardly a week passed that somebody did not come to my office at Chicago from somewhere, wanting to convey a piece of information to which I was not entitled. They usually did not ask me to conceal the fact that I came into possession of this information. All they asked was that I conceal from the Army the fact that they were the persons who had given it to me."

Szilard had been the first, in former days, to advocate that secrecy—of course only to a reasonable extent—should be maintained about scientific data. But he now became one of the first to be caught in the network of censorship regulations. A minor war flared up between Szilard and Groves, and it did not come to an end till long after the close of the Second World War. "Sure," Groves admitted, "we should never have had an atom bomb if Szilard had not shown such determination during the first years of the war. But as soon as we got going, so far as I was concerned he might just as well have walked the plank!"

Bohr found it particularly difficult to obey the secrecy regulations. After his escape from Denmark he was treated less as a man than as an extraordinarily valuable secret weapon which must on no account fall into the hands of the enemy. Accordingly, when this dangerously intelligent "consignment" was flown across the North Sea in a Mosquito, Bohr was given a seat just over the bomb bay, so that by turning a handle he could be dropped into the sea, a maneuver which would be carried out at once in the case of a German attack. Bohr arrived in London more dead than alive. For in his absorption in some problem of physics he had not heard the pilot tell him to put on his oxygen mask. Consequently, when the plane reached a great height, he fainted.

When Bohr, accompanied by his son Aage, landed in New York, two British detectives were with him. Unknown to the detectives, two secret-service men from the Manhattan Project organization joined the protective party, along with two officers of the Federal Bureau of Investigation. Bohr, the champion of freedom and straightforward dealing, was not pleased at being supervised by half a dozen watchdogs. He tried as often as he could to get away from them. It was really not easy for them to keep him unobtrusively in view, for he generally crossed the New York streets at prohibited places, obliging no fewer than six guardians of the law to join him in breaking traffic regulations.

Bohr could never get used to being called in the United States, on security grounds, by the name of "Nicholas Baker." Immediately after his escort had reminded him once more of this precaution he encountered, in the elevator of a skyscraper, the wife of his old colleague Halban. However, this lady had been divorced since their last meeting. "Are you not Frau von Halban?" Bohr inquired courteously. She retorted sharply: "No, you are mistaken. My name is Placzek now." But on turning to face him she exclaimed in astonishment: "But surely you are Professor Bohr!" He replied with a smile, placing his finger on his lips: "No, you are mistaken. My name is Baker now."

Before Bohr arrived for the first time in Los Alamos, Groves went to meet him. The General lectured him in the train, for twelve hours, on what he was and was not to say henceforth. Bohr kept on nodding. But once more Groves was disappointed. "Within five minutes after his arrival," the General reported later, in describing this prize specimen of his collection of "crackpots," "he was saying everything he promised he would not say."

The *enfant terrible* among the atomic scientists was the theoretical physicist Richard Feynman, as young as he was gifted. To enrage the censors he instructed his wife to send him letters to Los Alamos which were torn into hundreds of small pieces. The officials charged with the checking of correspondence were obliged to fit all the fragments of this jigsaw puzzle together again. It also afforded Feynman great amusement to work out the combination numbers of the steel safes in which the most important data of research were kept. In one case he actually succeeded, after weeks of study, in opening the main file cupboard at the records center in Los Alamos while the officer in charge of it was absent for a few minutes. Feynman contented himself, in the brief period during which he had all the atomic secrets at his disposal, with placing in the safe a scrap of paper on which he had written, "Guess who?" He was then able to feast his eyes on the horror of the security official as the latter perused the message which had found its way into the innermost sanctum of the Manhattan Project in some manner he was utterly unable to understand.

Groves could pardon Bohr his transgression of the sacred security regulations. He was even prepared to overlook Feynman's tricks, for they had the advantage, in his view, of keeping the security officials "on their toes." But the distinguished American scientist Edward U. Condon, one of the pioneers of experimental physics in the United States, excited a wrath in the General that remained implacable. Groves himself had invited Condon, in the summer of 1943, to act as his deputy in Los Alamos, working side by side with J. Robert Oppenheimer, then recently appointed head of the new bomb laboratory. Condon had acted as adviser to big industrial firms and possessed practical experience in dealing with questions of production, a qualification outside the province of the university-trained Oppenheimer. Simply because Condon did possess this experience he saw at once that compartmentalization was not practicable in Los Alamos without injury to the work. He therefore drew up rules of his own which demolished the artificial barriers erected between the separate departments. This step was regarded by Groves as downright insubordination. He caused Condon to be transferred elsewhere. The General considered that he could deal more easily with Oppenheimer in isolation, for his influence over the latter was uncommonly effective. The other atomic scientists could not understand at the time why this was so. It was not until much later that they found out the explanation.

Eight

The Rise of Oppenheimer (1939-1943)

At the time of Robert Oppenheimer's final appointment in July 1943 to the directorship of the Los Alamos laboratory he was in his fortieth year. For many people this year is the most important of their lives. It is then that they draw up their first real balance sheet. Then, perhaps, for the first time they seriously put to the judge who presides in the mind of each one of us the question: "How much of what I wanted to do in my youth have I actually accomplished? How much have I failed to do?"

Oppenheimer might well have been satisfied with what he had so far achieved. In the world of atomic research he was regarded as a theorist to be taken seriously. In university circles he was considered a particularly successful and popular teacher. The young man who had graduated with distinction under Max Born at Göttingen in 1927 found, when he returned home after two years' further study in Leyden and Zürich, that his high scientific reputation had preceded him. He was courted by a number of big universities. After some hesitation he decided to accept the offer of the University of California at Berkeley. When the dean of his faculty asked him what had caused him finally to decide in favor of them he answered, greatly to the other's astonishment: "Just a few old books. I was enchanted by the collection of sixteenth- and seventeenth-century French poetry in the university library."

Oppenheimer taught not only at Berkeley but also at the California Institute of Technology in Pasadena. As soon as he ended a course of lectures at the university most of his students would follow him, for the ensuing term, to this second seat of learning near Los Angeles. Despite his youth, "Oppie," as they called him, had come to be looked upon as a master and model by the rising generation of physicists in America, just as the great men of atomic research in Europe had been regarded by himself only a few years before. The veneration felt by the students for their hero was so great that consciously or unconsciously they imitated many of his personal peculiarities. They held their heads a little on one side just as he did. They coughed slightly and paused significantly between successive sentences. They held their hands in front of their lips when they spoke. Their ways of expressing themselves were often difficult to understand. They were fond of making obscure comparisons which sounded most pregnant and sometimes actually were so. Oppenheimer, himself a confirmed smoker, had the habit of clicking open his lighter and jumping up whenever anyone took out a cigarette or pipe. His students could be recognized from afar in the campus cafeterias of Berkeley and Pasadena by their custom of darting about from time to time, like marionettes on invisible strings, with tiny flames between their fingers.

Yet, unlike Rutherford, Bohr and Born, who had been both great teachers and great discoverers, Oppenheimer had not hitherto brought any epoch-making new ideas to light. He had certainly gathered a loyal circle of associates about him. But he had not yet established his own school of thought in physics. The many scientific papers he published in the periodicals of various countries unquestionably constituted valuable sections to the growing edifice of modern physics. But they laid no new foundations for it.

Oppenheimer's friends believed he was worried at not having climbed to the highest peaks of creative work in physics like Heisenberg, Dirac, Joliot and Fermi, who were about the same age as himself. His achievements might be considered exceptional by the academic world. But in his own more critical eyes he had not done enough. And since he was aware that, as past experience had shown, it was nearly always only the young, who were still capable of radical thinking, who hit on really new ideas, he was bound, as his fortieth year approached, to consider that he had failed to realize his highest hopes.

At this stage he was suddenly offered an opportunity to accomplish something exceptional in quite another direction. He was invited to take charge of the construction of the mightiest weapon of all time.

Oppenheimer had been thinking about the atom bomb ever since he had first heard, during a lecture by Bohr, of uranium fission and the vast quantities of energy it liberated. At a gathering in Washington in 1939 the Danish scientist had referred to the work of Hahn. But he had also drawn attention in particular to the deductions from it in the sphere of physics made by Frisch and Meitner. The information caused such a sensation that some physicists in the audience did not even wait to hear the end of the exposition but rushed straight off to their laboratories to reconstruct the experiments mentioned. A telegram containing a summary of Bohr's explanations was also sent to the physics department of the University of California. The German physicist Gentner, who was then working as a guest at the radiation laboratory at Berkeley, recollects that the very same day Oppie started on a rough calculation of the critical mass which could bring about explosion.

Nearly two years passed, however, before Oppenheimer was invited for the first time to participate in the secret initial studies of the uranium problem. In the autumn of 1941 he attended, at the request of the Nobel prize winner A. H. Compton, the two-day session of a special committee of the National Academy of Sciences, called to advise on the military application of atomic energy.

After this first contact with a complex of questions which were in the end to decide his destiny Oppenheimer at first returned to his teaching duties. But from that time on he could never rid his mind of the problems raised by the new weapon. In addition to his academic obligations he devoted much of his leisure to estimating how much uranium 235 might be necessary to cause atomic explosion. He was engaging in the same calculations Rudolf Peierls and his assistant Klaus Fuchs were carrying out at about the same time on the other side of the Atlantic, in England.

Oppenheimer also began, on his own initiative, to work in association with the Radiation Laboratory of his university. This group was directed by Ernest O. Lawrence, inventor of the cyclotron. They were experimenting with an electromagnetic method of separating uranium 235—in which a chain reaction can be brought about—from uranium 238, which is not susceptible of fission. Oppenheimer proceeded, with the aid of two students, to make a discovery which reduced the costs of this method by between 50 and 75 per cent.

Compton was so impressed by Oppenheimer's work, which no one had specially requested him to carry out, that at the beginning of 1942, when the American efforts to construct an atom bomb started on a grand scale, he asked Oppenheimer to give all his time to the project. In July of that year, during the summer vacation, Oppenheimer assumed the direction of a small group which discussed for a few exciting weeks the theoretical nature of the best kind of "FF (fast fission)-bomb." During these discussions, incidentally, mention was made for the first time, in definite terms, of a hydrogen bomb. But the question of its practical realization was shelved for the time being because it involved too many unknown factors.

Compton again expressed his great satisfaction with Oppenheimer's reports of progress. The latter's predecessor in charge of the group studying theory had been an excellent scientist but a bad organizer. "Under Oppenheimer," Compton recalls, "something really got done, and done at astonishing speed."

Oppenheimer came to the conclusion, in the course of his connection with the atom-bomb project, that the efforts of the various laboratories scattered all over the vast territory of the United States, as well as throughout England and Canada, would have to be concentrated at one particular spot; otherwise duplication of proceedings and confusion would be bound to result. He considered that a group of laboratories should be established at one place where collective work would be done under the direction of a single man by theoretical and experimental physicists, mathematicians, armament experts, specialists in radium chemistry and metallurgy, and technicians dealing with explosives and precision measurement.

Oppenheimer's idea found considerable support. As he was not only the spiritual father of the proposed new super laboratory but had also proved himself to be so outstanding as head of a team, Compton suggested that he should take charge of the experimental establishment in view, though it was not yet in being. It was in the autumn of 1942 that General Groves first became associated with Oppenheimer. In order to save as much as possible of the time of the extremely busy chief of the Manhattan Project, it was arranged that he and his two closest military collaborators, Colonels Nichols and Marshall, should meet the scientist in a reserved compartment of the luxurious train that ran regularly between Chicago and the West Coast.

In such cramped quarters, to the clatter of wheels over the rails, the first plans were laid for the new laboratory, the cradle of the still-unborn atom. The train thundered through the darkness of the night while man envisioned a light that would be brighter than a thousand suns.

Where was the new laboratory to be set up? The first place suggested was Oak Ridge, Tennessee, where a start had already been made, some months before, with the building of factories to produce explosives for the bombs. But that secret city lay dangerously near to the Atlantic coast, where German submarines were known to be in the habit of cruising and occasionally putting spies ashore. Two German agents had recently been picked up not far from Oak Ridge. They had not, incidentally, had the least intention of spying on the vast atomic establishments then in course of erection, but were only interested in contacting a certain German-American at the aluminum factory in Knoxville close by.

Nevertheless, this affair may have contributed to the decision of Washington to locate the second secret factory town for fissile material, the great plutonium works, far away at Hanford. Similar considerations led to the choice of a lonely region, which could be acquired comparatively cheaply, a long way from the Atlantic coast, for "Site Y," the provisional code name for the future birthplace of the atom bomb. Oppenheimer had at first suggested a site in California. But Groves, after inspecting the place, thought it unsuitable, too close to an inhabited district. The possibility had to be taken into account that preliminary experiments might lead to a premature explosion which would release dangerous radioactivity and imperil the civil population.

Oppenheimer then remembered a remote spot where he had once been educated as a boy, the rustic boarding school at Los Alamos, New Mexico. He had often in former days facetiously told his friends: "My two great loves are physics and New Mexico. It's a pity they can't be combined." But now it seemed, surprisingly enough, that this rather improbable combination might after all be brought about.

The Los Alamos Ranch School for Boys had been founded in 1918 by a retired officer named Alfred J. Connell. It stood at a height of over 7,000 feet above sea level, on a flat tableland, a mesa, forming part of the Pajarito (Little Bird) plateau of the Jemez Mountains. The scented pine woods and canyons of the region still swarmed, even after the First World War, with all sorts of winged and four-footed game. But the Indians who had once hunted there had long since abandoned their cave dwellings in the reddish-violet cliffs. They had migrated to villages of clay huts on lower ground. Only their "holy places" had been left behind on the mesas.

One of these consecrated spots of ground was also to be found on the Los Alamos mesa. When the founder of the Ranch School leased the surrounding land from its Indian owners he agreed not to build on the little piece of ground in question and to make no road through it. The area was protected by a low hedge. Probably in former times a kiva, the religious edifice of the Indians, had stood on it. One night in the autumn of 1942 some schoolboys, as a joke, threw a few empty food tins over the hedge. When Major Connell noticed them the next morning he had sinister forebodings. But for a while nothing occurred. Two or three weeks later, however, a car with a uniformed driver at the wheel came up the steep road to the mesa. As a rule only relatives of the boys or tradesmen took the trouble to visit Los Alamos. If tourists came they generally alighted and had a look round.

That car didn't stop. It drove slowly across the plateau and then turned. Groves, Oppenheimer and two of the General's adjutants were in it. "We didn't want to get out," Groves relates, "as otherwise we should have had to give some reason why we were inspecting the place. In any case, it was bitterly cold. I remember that detail very well, for nearly all the boys were wearing short pants and I thought they must be freezing. On the return journey I stopped the car a few times to see whether the very sharp bends in the road could be negotiated by heavy traffic. Then we drove back to Albuquerque, our starting point."

The utter isolation of the place appealed to Groves. He believed at the time that only about a hundred scientists and their families would be occupying the "Hill," with the addition, perhaps, in the course of time, of a few engineers and mechanics. He was not worried about the lack of residential accommodation—there was nothing available but the school buildings—by the difficult road up to the spot or the scanty supply of water. How wrong the infallible General's forecast was is proved by the fact that within a year of his first reconnaissance of Los Alamos 3,500 people were working and living there. A year later the figure had risen to 6,000.

Groves proceeded to take very rapid action. Under current wartime emergency legislation the headmaster of the Ranch School could do nothing to prevent the mesa and all its buildings being requisitioned for war purposes. He vacated Los Alamos, sent the boys home and cashed his indemnity check. Shortly afterwards he died, of a broken heart, as it was rumored in Santa Fe, the nearest town to Los Alamos.

On November 25, 1942, the Assistant Secretary of War, John McCloy, ordered the acquisition of Los Alamos. A few days later the first labor gangs arrived at the hill to excavate the ground for the foundations of the workshops of the "Technical Region." In March 1943 the first atomic scientists appeared. By the following June instruments scraped together from university laboratories had been hauled up the narrow road and new discoveries in nuclear physics immediately began to be made at Los Alamos.

As soon as Groves had decided on Oppenheimer's appointment, he found himself being criticized for it. "I was reproachfully told," he recalls, "that only a Nobel prize winner or at least a somewhat older man would be able to exercise sufficient authority over the many 'prima donnas' concerned. But I stuck to Oppenheimer and his success proved that I was right. No one else could have done what that man achieved."

The General was accustomed to demanding a good deal from his subordinates, but Oppenheimer threw himself into his task with such enthusiasm that even Groves was afraid he might overtax his strength. He had ordered the findings of the doctors, who had made a thorough examination of Oppenheimer, to be passed on to him, and knew he had been suffering from tuberculosis for years.

It seemed as though unsuspected physical resources had come to Oppenheimer's aid in those weeks. The first thing he had to do was to travel about the country by air or by rail to persuade other physicists to join him at the new secret laboratory on the edge of the desert. In the course of his recruitment tour he had first to dispose of the prejudices of many of his colleagues against Project S-1. During the two years and more it had taken people to make up their minds, while the atomic project stuck fast in a deadlock of overlapping authorities, the opinion had gone round among physicists that nothing good could ever come of the affair. In order to silence such doubts Oppie often went further than he should have gone on security grounds in his descriptions of the new studies and aims.

At that time he believed, in common with the most able of the specialists concerned, for example Hans Bethe, that the bomb could be ready within about a year. It is true that he could give no guarantee that the new weapon would be able to do its job. It might quite possibly turn out to be a dud. Nor did he conceal the fact that those who agreed to go to Los Alamos would have to sign, on security grounds, a more or less binding contract to remain there for the entire duration of the war. He added that they and their families would be cut off, as never before, from the outside world, and would be living in less than comfortable conditions.

In spite of Oppenheimer's frank admission of the many difficulties involved his recruiting campaign had an unexpectedly great success. His remarkable capacity for seeing the other point of view enabled him to find the right answer to the doubts expressed. Some physicists he terrified by the prospect of a German atom bomb. Others he attracted by his descriptions of the beauty of New Mexico. But to all he imparted the feeling of how exciting it would be to participate in the pioneering work to be carried out in this still-quite-novel field of research.

Probably, however, many of those he approached, who were mostly very young, agreed to his request for the reason, above all, that it was Oppenheimer who was to be their chief. His personal magnetism, hitherto only exercised upon his students, both male and female, now proved equally irresistible in wider circles. It was only seldom that one came across such inspiring personalities in the learned world. Oppenheimer bore no resemblance to the dry-as-dust specialist. He could quote Dante and Proust. He could refute objections by citing passages from the works of Indian sages which he had read in the original. And he seemed to be aflame with an inward spiritual passion. It would be extraordinarily stimulating to work in such close and intense association as would never have been possible in times of peace, with this and other outstanding atomic-research experts. Oppenheimer in fact possessed, as one of the victims subsequently put it, in irreverent but striking fashion, "intellectual sex appeal."

Accordingly, in the spring of 1943 some highly unusual tourists began slowly filtering into the sleepy city of Santa Fe, the former seat of the Spanish viceroys who had ruled Mexico centuries ago. The visitors did not take the usual interest in historical monuments or silver jewelry. And all of them appeared to be in a desperate hurry. They had been delayed, during their journeys from the eastern states or the Middle West, by the sudden intervention of troop movements, by missed connections or muddled flying schedules. They were consequently behind the times specified by their marching orders, instructions to report at No. 109 East Palace, Santa Fe. Thence they had been told that they would be transported to their secret destination, Site Y, thirty-five miles away.

The newcomers expected to be received in one of those prosaic office buildings generally occupied by officials throughout the world. But they found themselves instead, after reaching the address given them, standing at a wrought-iron gate, centuries old, which opened on to a small, picturesque courtyard of Spanish type. This situation was so unexpected and out of the ordinary that it acted on their senses like an enchantment.

At the other end of the patio, from which a subterranean passage had led to the governor's palace three hundred years before, a French window led to a rather small room. Here the new arrival would be welcomed like a long-lost son by motherly Dorothy McKibben. She would immediately set the nervous and overtired newcomer at his ease. Even the most distressed of the Ph——s (it was forbidden, on grounds of security, even to mention their telltale calling) soon began to relax, while the patient, good-natured Dorothy smilingly submitted to his questions. "Have my instruments arrived yet? We were told our furniture would be waiting for us. Is that right? Where shall I be living? What's that? No bus to Site Y before tomorrow morning? But I cut short a lecture so as to be here in good time!"

Mrs. McKibben had an answer ready for all such inquiries and comments. She had arranged for the new arrivals to be lodged for the time being in "guest ranches" in the neighborhood of Santa Fe, which had hitherto only been used by visitors on vacation, for up on the Hill, where room could only be found to house the delicate apparatus, no living quarters for the scientists were yet available. Mrs. McKibben took steps to track down mislaid luggage and lost children. The newcomers learned from her that in future their address would be simply "United States Army, Post Office Box 1663," and that either pseudonyms or numbers would henceforth be substituted for their names on their identity papers.

In particular, they were all strictly enjoined never to address one another by professional titles such as "Doctor" or "Professor" in Santa Fe, because the citizens might notice how many university men had all of a sudden come among them. When Teller arrived, he asked his colleague Allison, who had come to meet him, the name of the person represented by the statue in front of Santa Fe Cathedral. "It's *Archbishop* Lamy," Allison whispered. "But if anyone asks you who it is, you'd better say *Mr.* Lamy. Otherwise you'll have Dorothy on your trail right away."

"Although everyone was pressed for time, for there was not a day to be lost," Dorothy McKibben remembers, "it all began like a great, glorious and exciting adventure." She is still to be found, after the passage of some fifteen years and the explosion of some dozens of uranium and hydrogen bombs, sitting in the same office as before, surrounded by big framed photographs of those formerly under her wing. "They all look so very serious today," she remarks. "In those days they were not only younger but also more hopeful and enthusiastic. Probably it was not until much later that they realized the terrible seriousness of their undertaking."

Every morning buses drove up to the various ranches in order to collect the newly arrived experts and transport them to their work high up on the mesa of Los Alamos. The drivers sometimes forgot to call at one of the farms. Then the telephone at 109 East Palace would ring, to receive the answer: "Sorry! I hope we shall be able to find some sort of transport for you during the day. If not, I'm afraid you'll just have to go on practicing your riding for a while!" Even men who had never in their lives mounted a horse or put on riding breeches soon began to get used to living like pioneers in that remote, sunny region. While they waited for a normal existence to be organized on the hill they often cooked and ate out of doors, as if on a picnic. They made exploratory excursions with packhorses and tents to the canyons, in which some of the most secret testing installations were to be set up. The European scientists in particular, on those occasions, felt like characters in a Wild West novel. On Sundays they went for very long walks. Bethe and his wife even climbed some of the hitherto untouched peaks that surrounded Los Alamos.

Oppenheimer, especially, seemed to be in his element. He had often in former days spent his holidays, with his brother Frank, at a lonely farm not far from Los Alamos, where they both, at that time, used to live like cowboys. Oppie was now to be seen, if he did not happen to be away on one of his frequent tours, scrambling about over one of the various building sites. He would be sunburned as a native and clad in blue jeans, silver-studded belt and a garishly checked shirt. At Berkeley he had been in the habit of getting up as late as possible, insisting on starting no lecture earlier than eleven o'clock. But here in New Mexico he rose at daybreak. The indefatigable Oppie knew not only all the scientists, but also most of the laborers, by their first names. A decade and a half later they still used to ask Mrs. McKibben after "Señor" Oppenheimer. "Practically everyone," she says, "who took part in the building work at Los Alamos would have risked his life for him."

The holiday mood prevailing at Los Alamos during those weeks could even be noticed at the scientific meetings, where Oppenheimer as a rule took the chair. The rigid organization into divisions and groups was a later growth which had not yet begun at that time. It gradually arose during sessions of the special committees. In one of the more important initial discussions, it happened, for instance, that Edward Teller described the bomb's most intimate mechanism—two hemispheres brought into contact at a given moment, till the mass reached the critical point and exploded—in the form of a limerick. Like most verses of the kind, it would not only rhyme but be thoroughly naughty. Such was the carefree spirit in which work began on the most terrible of all weapons.

Nine

Fission of a Man (*1943*)

Before Oppenheimer, in 1942, became an official collaborator in the secret atomic project as a result of his appointment to the Metallurgical Lab, he was obliged, like all the rest, to fill in a long questionnaire. In this form he had acknowledged his membership in a number of left-wing organizations. His political interests had been aroused for the first time in 1933, after Hitler's accession to power, when some members of his family and some of his professional friends fell victims to the "German Revolution." Until then, like most of his scientific colleagues, he had troubled himself so little about events outside his technical, literary and philosophic preoccupations that he hardly ever read a newspaper or listened to the radio.

Oppenheimer's attention to politics increased as a result of the civil war in Spain and a personal encounter. In 1936 he had begun to pay court to a girl student of psychiatry named Jean Tatlock. Her father was Professor of English Literature at Berkeley. Jean was regarded as a devoted Communist. Through her Oppenheimer met some of the more prominent Communists in California. He began to read books about Soviet Russia and to reflect on the influence which political and economic events, such as the great Depression just coming to an end, might have upon human life.

In 1937 Oppenheimer inherited great wealth on the death of his father and began regularly to subscribe large sums in support of leftist causes. He also occasionally wrote brief anonymous pamphlets on the subject of contemporary events, printed them at his own expense and had them distributed by a group of anti-fascist intellectuals, including a number of Communists.

Oppenheimer states, with regard to his relations with Jean Tatlock: "We were at least twice close enough to marriage to think of ourselves as engaged." But in 1939, after the marriage had been several times planned and then again postponed, the scientist met a pretty brunette employed on experiments with fungi at the famous Plant Research Laboratory in Pasadena. Born Katharina Puening—and a relative of General Keitel —she had lived in Germany up to the age of fourteen and had just contracted a second marriage with an English medical doctor named Harrison. She and Oppenheimer fell so passionately in love that as soon as it could be arranged she freed herself from all her previous connections. The couple married in November 1940, ignoring the scandal thereby caused at Berkeley and Pasadena among the friends and relatives of their forsaken partners.

Oppenheimer dropped his associations with Communism almost at the same time as he terminated his relations with Jean Tatlock. Reports by his colleagues Placzek and Weisskopf at the end of 1938 on their experiences in Russia in the thirties had made a very deep impression on him. Both Oppenheimer and his wife—who also had a left-wing past—gradually tried to shake themselves free of acquaintances who remained loyal to the Party. In August the couple bought a house. That same year their first son, Peter, was born.

It proved difficult to make a radical break with the past. There were quite a number of people close to Oppenheimer who either sympathized with Communism or in some cases actually belonged to the Party. They included at least a few who had originally owed their interest in the ideas of the extreme left to Oppenheimer himself. Was he now simply to drop them? *

A similar situation prevailed in his private life. Jean Tatlock

* In contrast to this—as later stated by Oppenheimer himself—Haakon Chevalier told the author that to his knowledge Oppenheimer's association with Communist circles continued in fact much longer.

was still in love with him. After the cessation of their relations she had submitted to psychoanalytic treatment. Although herself a psychiatrist, she could not rid the innermost recesses of her mind from thoughts of Oppenheimer. Even after her marriage she wrote to him, visited him at his house, near her father's, and tried to reach him by telephone whenever she felt greatly excited or depressed. On certain occasions Oppenheimer consented to meet her, either out of pity, or a sense of guilt, or because he had himself not yet fully recovered from the wreck of a friendship that had lasted for years.

In June 1943 Oppenheimer escaped from the overwhelming burden of the duties imposed upon him in connection with the building work at Los Alamos and, in compliance with one of his former fiancée's urgent requests, went to see her at her house on Telegraph Hill in San Francisco. Late that afternoon they went out for drinks to the Top of the Mark, overlooking the city and the bay.

Oppenheimer had to tell Jean Tatlock that during the next few months, or possibly even for years, he would not be able to meet her; he was compelled to leave Berkeley for a time, with his wife and child. He added that he was forbidden to say what the nature of his official task was or to name the place to which he was being transferred.

Seven months after that last meeting Jean Tatlock took her own life.

The movements of Oppenheimer and his former fiancée in San Francisco on June 12 and 13 were all kept under continuous observation by agents of the counterintelligence branch of the Army, G2. They watched Oppenheimer accompany the young woman to her house in the evening. They knew that he spent the night there and that she drove him to the airport the following morning. The whole story was committed to paper down to the last detail and incorporated as "damaging information" in a comprehensive report. Ever since the end of May 1943 the scientist appointed to take charge of the atom-bomb laboratory had been the unsuspecting object of special investigation by the authorities. The security people did not trust him. They wanted to find out whether he might not still be keeping up his previous association, even now, with Communists. The visit to San Francisco at last gave Colonel Boris Pash, Deputy Chief of Staff of the G2 division in California, the "ammunition" he needed.

On June 29, 1943, Pash forwarded to the War Department in Washington a report summarizing the "results of supervision since arrival in San Francisco." In that report he expressly stated his suspicion that the "subject," as he invariably called Oppenheimer, in the usual style of police detectives, might be passing on to the Communists the scientific data obtained at Los Alamos before reporting them, if in fact he did so, to the government of the United States. This could very well be done through "contacts" like Jean Tatlock, who would then play her part by passing the information to the Party. Pash urged that every effort should be made to dismiss the "subject" at the earliest possible date and replace him by someone else.

The report was sent on to Groves in the middle of July with the intimation that for security reasons Oppenheimer's appointment to take charge of Los Alamos could not be confirmed. The General was dumbfounded. He was himself anything but sympathetic to Communists and was even then engaged, as he declared later, in doing all he could to prevent their infiltration into the Manhattan Project. As early as this, in fact, he supported the view that the country and the government trusted their Russian ally too implicitly.

And now it turned out that his closest, most indispensable collaborator was a Red! He sent for Oppenheimer. The latter immediately assured him that he had long since broken with the Communists.

Groves wondered whether he could believe in this change of Oppenheimer's opinions, but he felt bound to do so. Oppenheimer had only recently once more proved himself indispensable in dealing with the material difficulties which were threatening to damp enthusiasm at Los Alamos where the residential barracks erected on the mesa by units of the Army Pioneer

Corps proved cramped, uncomfortable and exposed to the risk of fire. The streets became by turns, according to the weather, full of dust or seas of mud. Only a single small group of pines had escaped the bulldozers. They had been rescued from the fate of the rest by a sit-down strike organized by Micci, Edward Teller's wife. When the men returned home tired after the day's work their wives, who for the most part were also obliged to do professional work, complained of the lack of domestic servants and unsatisfactory supplies of food. Oppenheimer managed to raise their spirits. He studied each individual problem himself, promised improved conditions and stressed the fact, in particular, that such small vexations, after all, counted for little in comparison with the importance of the job they were all doing.

Groves felt he could not do without this man, whether as scientist or organizer. He determined to keep him under his personal supervision. A War Department directive had ordered him not to lose a single day in producing the new weapon. Consequently he had been given unusual plenary powers, overriding all other decisions and regulations. He now determined for the first time to make use of this authority. On July 20, 1943, he sent the following telegram:

WAR DEPARTMENT, OFFICE OF
THE CHIEF OF ENGINEERS,
Washington, July 20, 1943

Subject: Julius Robert Oppenheimer.
To: The District Engineer, United States Engineer Office,
Manhattan District, Station F, New York, N.Y.

I. In accordance with my verbal directions of July 15, it is desired that clearance be issued for the employment of Julius Robert Oppenheimer without delay, irrespective of the information which you have concerning Mr. Oppenheimer. He is absolutely essential to the project.

L. R. GROVES, *Brigadier-General*,
CE.

This step seemed at first sight to settle the question of Oppenheimer's offending past. The General had cut the knot with a slash of his sword. Oppenheimer's gratitude knew no bounds. He had been reluctant all his life to put his whole heart into any single cause. His exceptional intelligence and clearsightedness had always shown him simultaneously the opposite of every view and the drawbacks of every undertaking. Consequently, despite his sympathy with Communism, he had never been a member of the Party. His great fear, more than anything else, was that he might become an accomplice and tool of a single one-sided idea, the politics of a single country; such a course invariably meant compromise, support for a principle which could not bear the light of absolute honesty and intellectual integrity.

Robert Oppenheimer, however, had now decided to devote himself entirely to the service of his native land. For the first time he believed that he had both feet on the ground of something real. That reality had been fashioned out of coarse material, no doubt. Rough, simple men like Leslie Groves had the biggest say in it, but they were ready to listen to the advice of a superior mind. Oppenheimer had come down from the rarefied air of the heights. He was no longer just an "unpractical" and "rootless" intellectual. Now he finally, finally, finally belonged. He had always shied away from "complicity"—a term he used frequently—more than anything else he had feared that he might become the tool.

It must have been as a result of such newly discovered patriotic sentiments that Oppenheimer paid a certain call at the end of August 1943, only a few weeks after Groves had taken his part with such impressive consequences. Happening to be at Berkeley, the scientist entered the office of the security agent, Lyle Johnson, established in one of the university lecture rooms. He intended to reveal a certain occurrence which he had kept a secret for months. The immediate cause of his visit was to talk about one of his former pupils, named Rossi Lomanitz, who had got into trouble. Oppenheimer had in the past persuaded the young man to join in the work, despite his conscientious scruples against participating in the production of the atom bomb. Lomanitz was now on the point of being expelled from the organization for pacifist and Communist propaganda. Oppenheimer proceeded to ask Johnson for permission, if that were not contrary to security regulations, to talk to Lomanitz and "bring him to his senses." This inquiry seems only to have been a pretext for Oppenheimer's visit, for in the course of further conversation he suddenly made certain astonishing statements. He had known for some time, he said, that the Russians were trying to obtain information about the American atom-bomb project. Oppenheimer added that an Englishman named George Eltenton, who had worked in the Soviet Union for five years before the war, had approached a certain personage, whose name he withheld, and asked the individual in question to get in touch, on his behalf, with physicists at work on the Manhattan Project.

Johnson listened attentively. He considered the information particularly important. For he himself, together with his immediate superiors, Colonels Pash and Lansdale, had been on the trail, ever since the end of February, of a Communist espionage organization which they assumed to be sending reports on the progress of American atomic armament. Three persons under suspicion in this connection, including the Lomanitz whose conduct was at present a subject of dispute, were pupils of Oppenheimer.

The counterintelligence branch still resented the fact that Groves had simply shrugged off their warnings about Oppenheimer, and did not believe for a moment that the man who had been put in charge of Los Alamos against their will had any patriotic motive in deciding to give them this belated information on the subject of attempted Soviet spying. They suspected that Oppenheimer had been told by his former pupils that their behavior was already under investigation. He had only "confessed," the security men supposed, in order to anticipate an an inquiry bound to be made, sooner or later, into his own activities.

The counterintelligence pretended, from that time on, to

regard him as a friendly witness. But they treated him, in reality, as if he were already in the dock, and made constant efforts to involve him in contradictions.

Johnson began by requesting Oppenheimer, with the greatest courtesy, to talk the whole matter over, once more, in detail, with his own chief, Colonel Pash. Boris Pash was the son of the Metropolitan of the Russian Orthodox Church in the United States. He was a huge fellow, who had only recently been appointed a specialist in "Communist infiltration." Prior to that appointment he had been a football coach at Hollywood High School. His Russian origin was enough to qualify him for this wartime post. Pash possessed a considerable measure of reckless daring which had got him into serious trouble. In order to teach certain officers to take greater care of military documents he instructed some of his subordinates to break into their quarters and make away with some highly confidential papers. A certain "big shot" took this proceeding in very bad part and very nearly had Pash dismissed. Pash, accordingly, had every reason at this time for wishing to prove himself a success.

When Oppenheimer first met him face to face, Pash already knew a good deal about him from the reports of his agents, as well as from the clandestine photographs and films they had taken. The Colonel had hidden microphones in his office before the interview began and had also placed a tape recorder in the next room. Every word of the long conversation that followed between the inquisitor and his witness was taken down without Oppenheimer's knowledge.

When Dostoevsky invents dialogue of this kind it is full of deep thought and brilliant phrasing. But how different it sounds in real life. The words seem utterly devoid of significance. They are intended to mean something, certainly. But at the same time they serve to veil that meaning. There is a great deal of mere chatter, to avoid having to come to the point. There is a great deal of stammering and hesitation, for neither side wants to state the plain truth.

The conversation began with an exchange of ordinary courtesies.

Pash: This is a pleasure, because I am interested to a certain extent in activities and I feel I have a certain responsibility in a child which I don't know anything about. General Groves has more or less, I feel, placed a certain responsibility in me and it's like having a child, that you can't see, by remote control.

I don't mean to take much of your time——

Oppenheimer: That's perfectly all right. Whatever time you choose.

Pash: Mr. Johnson told me about the little incident, or conversation, taking place yesterday in which I am very much interested and it had me worried all day yesterday since he called me.

Oppenheimer behaved at first, at this stage, as though he did not understand what Pash wanted to see him about. He began again to tell the story of the difficulties his pupil Lomanitz was having with the authorities. But Pash at once steered the conversation round to the subject which Oppenheimer wanted to avoid, that of his disclosures about the alleged Soviet attempts at espionage. Pash wanted to know the name of the intermediary approached by Eltenton. Oppenheimer did not answer that question. Instead of doing so he went on to say, evidently irritated and hoping to divert Pash's attention, that the unknown intermediary referred to had already been talking to three atomic scientists.

Pash: Yes. Here's the thing—we of course assume that the people who bring this information to you are 100 per cent with you and therefore there is no question about their intentions. However, if——

Oppenheimer: Well, I'll tell you one thing—I have known of two or three cases, and I think two of the men were with me at Los Alamos—they are men who are very closely associated with me.

Pash: Have they told you that either they thought they were contacted for that purpose or they were actually contacted for that purpose?

Oppenheimer: They told me they were contacted for that purpose.

Pash: For that purpose.

Oppenheimer: That is, let me give you the background. The background was—well, you know how difficult it is with the relations between these two allies, and there are a lot of people who don't feel very friendly to Russia, so that the information—a lot of our secret information, our radar and so on, doesn't get to them, and they are battling for their lives and they would like to have an idea of what is going on and this is just to make up, in other words, for the defects of our official communication. That is the form in which it was presented.

Pash: Oh, I see.

But this further information did not produce the effect Oppenheimer probably desired. His explanation merely increased Pash's interest in the unknown intermediary. He repeatedly brought the conversation back to this point.

Pash: Well, now I may be getting back to a little systematic picture. . . . These people whom you mentioned, two are down there with you now. Were they contacted by Eltenton direct?

Oppenheimer: No.

Pash: Through another party?

Oppenheimer: Yes.

Pash: Well now, could we know through whom that contact was made?

Oppenheimer: I think it would be a mistake, that is, I think I have told you where the initiative came from and that the other things were almost purely accident and that it would involve people who ought not to be involved in this.

Oppenheimer declined to allow Pash to wheedle any more out of him. He steadily refused to name the person who had taken the message, though this was what Pash obviously longed to know. But he did assure the agent that in Los Alamos at any rate, where Oppenheimer himself was in charge, no Communist agitation or espionage was to be feared. In a slightly melodramatic phrase, which smacked more of the official than of the man he had hitherto been, the Director of the Los Alamos establishment solemnly declared:

"If everything were not proceeding according to plan and in due order down there I shouldn't object in the slightest to being shot."

But the counterintelligence branch was not satisfied with such rhetorical protestations of loyalty. Oppenheimer's refusal to take another step along the path of denunciation, which he had already entered, made him all the more suspect. Ten days after his interview with Oppenheimer, Pash sent to his chief at the Pentagon, Colonel Lansdale, the following description of the judgment he had formed of the scientist:

"This office is still of the opinion that Oppenheimer is not to be fully trusted and that his loyalty to the Nation is divided. It is believed that the only undivided loyalty that he can give is to science and it is strongly felt that if in his position the Soviet government could offer more for the advancement of his scientific cause he would select that government as the one to which he would express his loyalty."

The suspicion of the intelligence agents that Oppenheimer had not so far been completely frank in his conversations with them was well founded. He had been lying. He had been hiding something from them. They assumed that the "something" in question concerned the connection which they supposed him to be still maintaining with the Communist Party or even with the Soviet espionage organization. But Oppenheimer had in actual fact broken off his former erratic relations with Communism. What he most feared in present circumstances was that the authorities might dismiss him in spite of this fact, if they gradually obtained more and more information about his left-wing activities in the past. For he would then lose his important new appointment and be driven back into no man's land.

But the men in the counterintelligence branch were not to be put off by mere hints. They wanted to know the whole truth. That, however, was just what Oppenheimer, in his mental condition at that time, could not tell them. For its chief result would have been to incriminate himself and thus once more seriously endanger his position as Director at Los Alamos.

In point of fact only one scientist, not three, had been approached by the unknown intermediary. And the name of that scientist was Robert Oppenheimer.

What had actually happened was that towards the end of the year 1942 or not long after the beginning of 1943—the precise date has never been ascertained—the Oppenheimers, who were then living at their house, "Eagle Hill," in Berkeley, were visited by a neighboring married couple, the Chevaliers.

Oppenheimer had known Haakon Chevalier, a lecturer in Romance languages at the University of California, since 1938. The atomic physicist had very soon struck up a genuine friendship with this colleague from another faculty and two years older than himself. Their amicable relations arose in spite or possibly just because of the fact that Chevalier was so entirely different from Oppenheimer. The tall, broad-shouldered man, with his Norwegian Christian name and French surname, radiated so much simple enthusiasm and cordiality that Oppenheimer confided more in Chevalier than he did in anyone else. Oppenheimer could argue and speculate for hours on end with physicists like Robert Serber and Philip Morrison, who were his personal friends. But with Chevalier he felt that he could either remain silent or talk, with some slight degree of nostalgia, about distant Europe and its poets.

Chevalier himself had been born in the little town of Lakewood, New Jersey, but he had returned with his parents to France, his father's country, when he was two years old. Later he went to Norway, from which his mother came. Many people there still remembered his grandfather, a corn merchant, and friend of Grieg and Ibsen. In 1914, on the outbreak of war in Europe, the family had returned to the United States. When the war ended Haakon was a dreamy youth of eighteen, a poet and also something of a wanderer like Knut Hamsun. From love of adventure and a desire to see the world he had gone to sea. It was not until he had served "one year before the mast" that the young man, still avid for experience, had sat down again in the schoolroom and soon proved himself a brilliant expert on French literature. With Haakon Chevalier it was possible for Oppenheimer to escape from physics, to discuss Anatole France or Proust, his favorite author, or else simply to try out recipes for exotic and highly seasoned dishes which the two men prepared together in Oppenheimer's kitchen.

Such associations are often broken when one of two friends marries. But in this case the relationship became even closer after marriage. Chevalier and his wife, Barbara, were among the few people who stood by Robert and Katie Oppenheimer when in November 1940 the marriage between these two caused so much gossip and excitement.

As a bachelor Oppenheimer had lived in two rooms opening on to a broad terrace. It was generally extremely cold and uncomfortable there. He used to leave the windows open day and night because of the tuberculosis from which he had suffered for years. But on his marriage he began to look around for a home of his own. Chevalier was of assistance to Oppenheimer in that connection. He himself lived in an old English country house which had been transported in its entirety to the distant California coast for the San Francisco World's Fair in 1915. This museum piece had been bought, when the exhibition closed, by a lady who had transferred it, with immense toil and trouble, to the top of a hill overlooking Berkeley. She also had a second house close by, a long, white stucco building of Spanish type, with a spacious and cozy living room that had a painted wooden ceiling, a tiled floor of brownish-red earthenware and a large fireplace. A steep road led up to the house, perched, like an eagle's nest, on the brink of a precipice.

This dwelling on Eagle Hill, surrounded by tall cypresses, became one evening the scene of a certain conversation between Robert Oppenheimer and Haakon Chevalier destined to exercise a fateful influence on the future lives of them both. At the time they considered the conversation so unimportant that neither could afterwards remember exactly what words were used. While the two wives were talking together in the living room, Chevalier had followed his host into the small kitchen which adjoined it. Oppie began to mix Martinis. Chevalier proceeded to inform him that he had recently been talking to a man they both knew named George Eltenton. Eltenton had complained to him that in spite of the fact that the American and Soviet governments were allied no interchanges of new scientific information took place between the scientists of the two countries. He went on to ask Chevalier whether it would not be possible to persuade Oppenheimer to pass on scientific data in a private capacity. Oppenheimer reacted to Eltenton's suggestion in the manner Chevalier had foreseen.

So far as Chevalier remembers Oppenheimer exclaimed: "That's not the way to do these things!" According to Oppenheimer's own subsequent statement his answer was even more pointed. He believed he had retorted: "That would be a frightful thing to do!" and "But that would be high treason!"

This was the end of the conversation. Since the two men were in full agreement on the subject it was never again raised between them. They returned to the big room and drank their cocktails.

But while the Chevaliers were on their way home that evening Haakon's wife remarked: "I don't know why, but somehow I don't trust Oppie."

It was a presentiment, nothing more. Chevalier took no notice of his wife's warning at the time.

Counterintelligence continued to harry the atomic scientist. As Boris Pash had not succeeded in learning anything more definite about the unknown intermediary or the three scientists he had allegedly approached for information, Oppenheimer was summoned to Washington. The authorities hoped that a more adroit inquisitor than the somewhat clumsy Pash might contrive to worm the secret of the middleman's identity out of him. On September 12, 1943, a fresh cross-examination of Oppenheimer began in an office at the Pentagon, conducted on this occasion by Colonel John Lansdale, Jr., himself, the able Chief Security Officer for the entire atomic project, then only just thirty-one years of age. Once more preliminary measures were taken to ensure registration of the conversation by means of a concealed microphone connected with a tape recorder. Lansdale proceeded to show a good deal of ingenuity in his efforts to crack open Oppenheimer's secret. He must have received the impression from a previous encounter with Oppenheimer at Los Alamos, which had taken place exactly a month before and was the first time they had met, that the other's defenses could be broken down by flattery. Accordingly, Lansdale immediately made for this weak spot.

Lansdale: Well, now, I want to say this—and without intent of flattery or complimenting or anything else, that you're probably the most intelligent man I ever met and I'm not sold on myself that I kid you sometimes, see? And I'll admit freely that at the time we had our discussion at Los Alamos I was not perfectly frank with you. My reasons for not being are immaterial now. Since your discussion with Colonel Pash I think that the only sensible thing is to be as frank with you as I can. I'm not going to mention certain names, but I think that you can give us an enormous amount of help, and as I talk you will realize, I think, some of the difficulties that have beset us.

Oppenheimer: There are some, I think, that I know already.

Lansdale: That's right. Now, I will say this, that we have not been, I might say, asleep at the switch, to a dangerous extent. We did miss some things, but we have known since February that several people were transmitting information about this project to the Soviet government.

Oppenheimer: I might say that I have not known that. I knew of this one attempt to obtain information which was earlier, or I don't, I can't remember the date, though I've tried.

Lansdale: Now, we have taken no action yet, except with respect to Lomanitz.

Oppenheimer: Are they people who would be in a position to transmit substantial information?

Lansdale: Yes, I'm so informed, I don't know personally, of course.

Oppenheimer: Well, Lomanitz by virtue of being a theoretical physicist would probably have a rather broad knowledge of the things he is working on.

The hearing thus began to the advantage of the questioner. Four weeks previously Oppenheimer had still been trying to protect and defend his pupil Lomanitz. But now he had ceased to do so and even seemed ready to give information which might damage him. Nevertheless, these promising signs proved to be deceptive. As soon as Lansdale approached the topic of the Eltenton affair, he was met by the same blank wall that

Pash had previously found insurmountable. Lansdale, accordingly, started dropping his ballast overboard. He said he would be prepared to remain in ignorance of the names of the three scientists. But he added that he absolutely must have the name of the intermediary, so as to be able to prevent similar attempts at contact in the future. But even then he could not convince Oppenheimer of the necessity of this disclosure.

Oppenheimer: I've thought about it a good deal because Pash and Groves both asked me for the name, and I feel that I should not give it. I don't mean that I don't hope that if he's still operating that you will find it. I devoutly do. But I would just bet dollars to doughnuts that he isn't still operating.

Lansdale: I don't see how you can have any hesitancy in disclosing the name of the man who has actually been engaged in an attempt at espionage to a foreign power in time of war. I mean, my mind just doesn't run along those channels, and——

Oppenheimer: I know, it's a tough problem, and I'm worried about it a lot.

Lansdale: I can understand personal loyalty, yet you say he's not a close friend of yours. May I ask, do you know him as a Communist?

Oppenheimer: I know him as a fellow traveler.

Lansdale called twice more upon Oppenheimer to reveal the name of the mysterious unknown middleman. Twice more the request was refused. The refusal was all the more astonishing on account of the thoroughly co-operative attitude of the man under examination with respect to other persons about whom Lansdale asked him. Oppenheimer neither made any secret of the Communist sympathies of the wife of his friend Robert Serber nor did he flatly decline to answer another suggestion put to him by the counterintelligence agent in the following terms:

Lansdale: Could you get information about who is and who isn't a member of the Party?

Oppenheimer: I don't know whether I could now. At one time I could have. I never tried to.

Lansdale: Would you be willing to?

Oppenheimer: Not in writing, I think that would make a very bad impression.

Lansdale: No, not in writing.

Oppenheimer: I don't know anyone at Los Alamos who could give information of that kind. I could get partial information.

Lansdale tried one last trick. After assuring Oppenheimer, "Don't think it's the last time I'm going to ask you, because it isn't," he continued:

Lansdale: Well, I want to say that personally I like you very much and I wish you'd stop being so formal and calling me Colonel, 'cause I haven't had it long enough to get used to it.

Oppenheimer: I remember at first you were a captain, I think.

Lansdale: And it hasn't been so long since I was a first lieutenant and I wish I could get out of the Army and back to practicing law, where I don't have these troubles.

Oppenheimer: You've got a very mean job and——

Lansdale: I want you to know that I like you personally and believe me it's so. I have no suspicions whatsoever and I don't want you to feel that I have and——

Oppenheimer: Well, I know where I stand on these things. At least I'm not worried about that. It is, however, as you have asked me, a question of some past loyalties. . . . I would regard it as a low trick to involve someone where I would bet dollars to doughnuts he wasn't involved.

Lansdale: O.K., sir.

That "O.K." did not mean the matter was settled. Lansdale did not make the slightest reference, in the memorandum to Groves which he drew up after the hearing, to the personal liking for and confidence in Oppenheimer which he had expressed so volubly at the interview. On the contrary he urged that greater pressure should be brought to bear on the scientist with the object of forcing him at all costs to disclose the name required.

The papers relating to the case which lay on General Groves's desk included an exhaustive critical study of Oppenheimer's character by the intelligence agent Peer de Silva. It culminated in a remarkable suggestion. In September 1943 de Silva had written:

> It is considered that Oppenheimer is deeply concerned with gaining a worldwide reputation as a scientist and a place in history as a result of the DSM project. It is also believed that the Army is in the position of being able to allow him to do so or to destroy his name, reputation and career, if it should choose to do so. Such a possibility, if strongly presented to him, would possibly give him a different view of his position with respect to the Army, which has been, heretofore, one in which he has been dominant because of his supposed essentiality.

The head of the Manhattan Project acted in precise conformity with this proposal when some weeks later, in December 1943, he himself examined Oppenheimer, in private. He frankly told the Los Alamos chief that he would now have to order him to reveal the name which he had been keeping secret unless Oppenheimer himself chose, at this late stage, to reveal it of his own accord. The latter had already informed the General, two months before: "General, if you order me to tell this, I will tell you." At the time Groves had not insisted. But now he had decided not to wait any longer.

In this situation Oppenheimer could have taken up the attitude that his duties at Los Alamos were purely scientific and that he was not bound to act as an informer on behalf of counterintelligence or even to obey orders as a soldier. If the authorities refused to concede this point he could always have resigned. Fortunately, there was no such thing in the United States as torture and the arrest of relatives, such as was practiced in the totalitarian countries in dealing with obstinate persons. Consequently, if Oppenheimer had really wished to do so he could have perfectly well gone on withholding all information about Chevalier, of whose innocence he was convinced. But he gave in, and at last disclosed the name of the man whom he had himself, as he admitted later, so deeply incriminated by his exaggeration of what had in fact occurred. He thus finally did what he had been vainly urged to do for months, so saving himself and his career. Within a very short time that career was to take him to the pinnacles of fame and power.

No one except Oppenheimer himself and the officials dealing with his case had any idea at that time, in the middle of the war, of the personal trial which the Los Alamos Director had undergone. Chevalier himself knew nothing of it, though soon afterwards, without realizing that his friend had betrayed him, he was dismissed from his tutorial post for unknown reasons. It was not until more than ten years later, after he had been driven into exile and was still without an appointment, that he finally learned who had informed against him and thus brought his academic career to an end.*

* In 1954 Oppenheimer admitted, during an official hearing, that his stories about the mysterious intermediary, whom he subsequently named to Groves as Chevalier, had been "idiotic" and "a tissue of lies."

Ten

The Pursuit of Brains (*1944-1945*)

In December 1942 the atomic scientists working in the Metallurgical Laboratory at the University of Chicago heard that Hitler was going to risk his first air raid on the United States on Christmas Day. The rumor stated, moreover, that the object of the attack would be the great city itself, with its population of millions. Chicago was at that time the center of American atomic research. The whisper went round that the Germans would probably drop, not the usual bombs, but great quantities of radioactive dust to poison the air and water of the city. This story, noted down by Sam Goudsmit, was so generally believed that certain physicists sent their families into the country, while military posts distributed Geiger counters.

It was no accident that such tales began to circulate the very week after the first primitive uranium reactor, designed by Fermi, had been started in the windowless cellars below the grandstands of Stagg Field. The practical realization of a controlled chain reaction in the uranium pile was a brilliant scientific achievement. The rumors about radioactive "death dust" were the dark shadows cast by that event. In future it would be possible to produce artificially in the uranium-burning ovens now proved to be technically feasible, tons of dangerous radioactive matter which only occurred in microscopic quantities in nature.

If a uranium reactor of this kind had now at last been produced in Chicago, after the American atomic project had got off to such a slow start, it must also have long since been constructed somewhere in Germany, the Allied nuclear physicists believed. They had never ceased to live in terror of Hitler's advantage in the atomic-armaments race. It was to be assumed that the Germans had already enough radioactive matter in their piles to poison all the large cities of their enemies.

As a precaution against these and other surprises to be expected from the German armament laboratories the American high command instituted in the autumn of 1943 a special intelligence unit, to be disembarked with the first troops invading Europe, for the purpose of collecting specific information on the state of German atomic armament. This top-secret special division was given the somewhat transparent code name of "Alsos." The word is a literal translation into Greek of the name "Groves." Its members, contrary to all the principles of espionage, were actually rendered recognizable by a unique badge, consisting of the alpha sign in white, transfixed by red forked lightning to symbolize atomic power.

In November 1943 Colonel Boris Pash was appointed to command Alsos. He was at last able to hand over the unsatisfactory and obscure "Oppenheimer affair" to others and devote himself to more interesting tasks. The first discoveries made by Pash in Europe after ransacking documents at the University of Naples were so few and far between that he was recalled. It was decided that he should share his command on his next mission with an atomic-research scientist. Such a team would probably acquire more interesting information. The choice fell on Samuel A. Goudsmit, the well-known Dutch experimental physicist, whose hobby had long been the study of the latest methods of criminal investigation.

Goudsmit was working on the radar project of the Massachusetts Institute of Technology. He had no idea why he had been selected for the mission in question. Later on, while examining a file dealing with prospective candidates for appointment to the Alsos mission, he came across a confidential evaluation of his own character, which had got there by mistake. It simply stated that he had "certain valuable qualifications" for the post and "certain disadvantages." Goudsmit observes that he immediately guessed what the disadvantages were and adds that the advantages were probably that in the first place, although a nuclear physicist, he had not yet been employed on the Manhattan Project. Consequently, if he happened to be captured by the enemy during the fighting in Germany, no important atomic secrets could be extracted from him. In the second place he spoke fluent French and German. He had worked in Leyden under Ehrenfest, a pupil of Bohr, and then, during the twenties, for some time at Bohr's Institute in Copenhagen. There, while still a very young man and before obtaining his doctor's degree, he had made one of the most important discoveries of modern physics, the so-called "spin" of electrons.

Goudsmit had been nicknamed "Uncle Sam" in the family circle of atomic experts, though he was not so Americanized as all that, despite his residence since 1927 in the United States. He was more cheerful and genial and far more versatile than the other physicists. In addition to his passion for criminology, he was a first-rate Egyptologist, collected scarabaei and told stories brilliantly. But he was above all a warm-hearted and modest man, revered by his pupils and loved by his friends.

"Many physicists only have high-voltage current in their veins," said someone who had known him a long time. "But Sam has blood. He knows that the world contains other interesting things besides equations and cyclotrons." It was typical of Goudsmit that he gave the following advice to a young nuclear physicst who wanted to go to an atom-bomb test in Nevada. "If you want to see a show, why don't you buy a ticket to one of the current Broadway musicals? That might help you more with your work than going west. Pauli won the Nobel prize by going to the theater, you know. He was watching a revue in Copenhagen when the exclusion principle came to him." *

* Pauli, in a letter to the author, tries to deflate that favorite story of the brilliant "raconteur" Goudsmit by stating: "The idea came to me during a simple promenade. . . ."

Colonel Pash, military head of the Alsos mission, accompanied the very first Allied troops that entered Paris at the end of August 1944. Only two days later Goudsmit and his scientific personnel followed. As civilians they were obliged to keep a little behind the front-line troops. Their first objective was the occupation of the Collège de France, where Joliot-Curie's laboratory was situated. He had not left France since the German invasion. Many Frenchmen considered him, in those days, a *collaborateur*. He was regarded as a traitor for having handed over his laboratory intact to the Germans, in 1940. But in reality this apparent capitulation merely served to camouflage the scientist's extremely active participation in the French resistance movement. After the departure of Wolfgang Gentner the laboratory became an arsenal for the Paris *maquis,* although—or perhaps even because—other buildings in the group comprising the Collège de France were used as offices by the German Military Government. Joliot's quarters were never searched, for the simple reason that no one dreamed the scientist capable of such mad audacity. He had himself taken part in the last few days of street fighting for the liberation of the city. The man who had discovered, through his studies of neutron emission and chain reaction, some of the most important of the necessary preconditions for construction of the atom bomb used the most primitive form of bomb imaginable in defense of the barricades—ordinary beer bottles filled with gasoline and fitted with fuses.

Joliot was unable to supply any interesting information about the German atom bomb. Moreover, the authorities in Washington advised the greatest caution in dealing with him, for, a week after the liberation of Paris, Joliot declared that during the war he had transferred his allegiance from the Social Democrats to the Communists.

The Allied armies marched on to the invasion of Germany. It was hoped that they would very soon occupy Strasbourg, where, it was understood from Alsatians, several laboratories attached to the University were engaged on atomic research. The advance was held up, but the Alsos mission did not remain inactive. One of its members, Captain Robert Blake, accompanied the daring dash of a detachment in Holland, the first to reach the Rhine. He ventured into the middle of the river under heavy fire and filled some receptacles with the gray-green water. The bottles were sent back by special messenger to the Paris headquarters of the Alsos group and thence by the quickest possible route to Washington. It was believed that the Germans, if they possessed a uranium burner, would probably have had to use for cooling purposes water from a river flowing for a short part of its course through the "pile." The Americans themselves had used the Columbia River for their plutonium pile at Hanford. If that were so, it would be possible for chemical analysis to detect traces of radioactivity in the water taken from the river and thus enable the Alsos organization to get on the trail of the German project. The major who arranged the forwarding of the water from the Rhine to Washington facetiously added to the consignment a bottle of the best red wine from Roussillon for unofficial sampling, writing on the label: "Test this for activity too!"

Within a week a cable in code, addressed to the Alsos mission, was received from General Groves's office. It read: "Water negative. Wine shows activity. Send more. Action." The people in Paris laughed, telling one another: "Well, they must have liked the stuff!" No one dreamed that the cable meant anything more than an amicable participation in the major's little joke. But soon another cable followed the first. "Where are those wine bottles?" it demanded in deadly earnest. A secret German laboratory somewhere near the famous French vineyards was suspected and the matter had to be investigated at once. Consequently, the people in Washington had evidently not understood the joke. They must have poured that excellent Roussillon into test tubes and adulterated it with chemicals instead of drinking it.

At that particular moment Goudsmit was reluctant to send any of his collaborators on a wild goose chase into the vineyards of southern France. But all his attempts to convince Washington that it had misunderstood a simple, harmless jest came to nothing. The Pentagon insisted on its orders being carried out. Accordingly, Major Russel A. Fisher and Captain Walter Ryan were sent to Roussillon on special service. Before they started they were grimly warned by Goudsmit: "Do a complete job. Don't be stingy with the confidential funds. And above all be sure that for every bottle of wine you locate you secure a file copy for our office in Paris."

The two intelligence agents were taken by the French wine growers for scouts acting on behalf of American export firms. Whenever they made inquiries for radioactive Roussillon they received enthusiastic hospitality. They spent ten hilarious days. Then they returned to Paris with several basketfuls of red wine, grapes and samples of soil.

The agreeable experiences of Goudsmit, the physicist whom war had turned into a scientific intelligence agent, were in the minority. Wherever he turned he found the traces of distress and death in the scientific sphere as elsewhere. Great numbers of distinguished men of science had been imprisoned by the Nazis or deported. A typical case was that of the French physicist Georges Bruhat. His pupil Claude Roussel had hidden certain American pilots, who had been shot down, near the École Normale Supérieure. When the Gestapo began to suspect Roussel, Bruhat declined to betray his pupil and was punished by being sent to Buchenwald. There he continued to lecture to his fellow prisoners on astronomy until at last he died of starvation.

An even worse fate overtook the Alsatian Holweck, who had invented a new kind of machine gun, with a specially rapid action, for the French Army. He was tortured to death by the Gestapo in the attempt to force him to reveal the secret of his invention.

The case of two Dutch physicists, which gave Goudsmit a difficult problem of conscience to solve, was of a different order. They had both escaped to England during the war and performed valuable services on behalf of the exiled Dutch government. The head of the Alsos mission found proof, in captured German documents, that before making their escape, both had worked, to keep their families alive, for the benefit of the German armament industry. Ought he to, could he possibly, report this political offense by people who had once been his compatriots? He decided not to do so.

Finally, Goudsmit had one experience of a highly personal nature. Immediately after the liberation of Holland he had hurried to the Hague in the hope of finding some trace of his parents there. He had not heard from them since March 1943.

It was a melancholy homecoming. "The house was still standing," he recalls. "But as I drew near to it I noticed that all the windows were gone. Parking my jeep around the corner so as to avoid attention I climbed through one of the empty windows. . . . Climbing into the little room where I had spent so many hours of my life I found a few scattered papers, among them my high-school report cards that my parents had saved so carefully through all these years. If I closed my eyes I could see the house as it used to look thirty years ago. Here was the glassed-in porch which was my mother's favorite breakfast nook. There was the corner where the piano always stood. Over there had been my bookcase. What had happened to the many books I had left behind? The little garden in back of the house looked sadly neglected. Only the lilac tree was still standing. As I stood there in that wreck that had once been my home I was gripped by that shattering emotion all of us have felt who have lost family and relatives and friends at the hands of the murderous Nazis—a terrible feeling of guilt. Maybe I could have saved them. After all, my parents already had their American visas. . . . If I had hurried a little more, if I had not put off one visit to the Immigration Office for one week, if I had written those necessary letters a little faster, surely I could have rescued them from the Nazis in time."

Not long afterwards Goudsmit made a second shocking discovery. In the course of his search for documents dealing with the German uranium project he came upon a list of those condemned to death by the SS. It included the names of his parents. "And that is why," he writes, "I know the precise date my father and my blind mother were put to death in the gas chamber. It was my father's seventieth birthday."

On November 15, 1944, Strasbourg fell to General Patton. Once more Colonel Pash was among the first troops to march into the town. He and his special detachment occupied the Physics Institute forming part of the medical faculty of the University. A great many documents were found and four German physicists were captured. When Goudsmit came to cross-examine them, he felt, he says, somewhat embarrassed. After all, they were colleagues, whom Pash, to prevent any kind of collusion, had placed in separate cells of the town jail. "Yet I felt uncertain of myself," Goudsmit writes, "and even somewhat embarrassed in their presence, especially at the prospect of calling on a colleague in jail. How was I certain he deserved to be in jail? Or was this just normal in war?" The situation was so painful to him that he did not immediately tell the German scientists that he was a physicist. The prisoners, for their part, declined to make any statement; they had no intention of revealing anything about their work to an enemy. Goudsmit had never realized so clearly before as he did in that Strasbourg prison what the war had done to science and scientists and how fundamentally different, in fact quite antagonistic, the rules were which governed scientific life and the practice of warfare. On the one side frankness and international friendship prevailed, on the other secrecy and compulsion.

Goudsmit had hoped to capture Weizsäcker, who had recently been made Professor of Theoretical Physics at Strasbourg. He had been away from the University for the last three months, but he had left many of his papers behind. Goudsmit and one of his assistants sat far into the night, by candlelight, over these letters and documents. The hollow booming of artillery fire from the other side of the Rhine and the muttered exclamations of the G.I.'s who were playing cards in the same room formed a background accompaniment to the exertions of the two scientific detectives, as they searched among the hints and casual remarks in Weizsäcker's correspondence for clues to the state of German atomic research. Suddenly and almost simultaneously they both uttered a triumphant shout. There it was, the thing they had been looking for month after month! A whole bundle of papers dealing with the German uranium project had come to light!

It was unequivocally clear from the papers found by Goudsmit in Weizsäcker's Strasbourg office that the Germans, who had always been supposed to be ahead in atomic research, were in fact at least two years behind the Allies in this field. They had no factories yet for the production of the U235 or PU239 (plutonium) needed for chain reaction in the bomb. Nor had they any uranium burners comparable with the American apparatus for that purpose.

The date of the turning point in German atomic research had been June 6, 1942. On that day Heisenberg had given Speer, the Minister of Supply, and his staff an account of the situation with regard to his own studies. Heinsenberg himself reports: "Definite proof had been obtained that the technical utilization of atomic energy in a uranium pile was possible. Moreover, it was to be expected on theoretical grounds that an explosive for atomic bombs could be produced in such a pile. Investigation of the technical sides of the atomic-bomb problem—for example, of the so-called critical size—was, however, not undertaken. More weight was given to the fact that the energy developed in a uranium pile could be used as a prime mover, since this aim appeared to be capable of achievement more easily and with less outlay. . . . Following this meeting, which was decisive for the future of the project, Speer ruled that the work was to go forward as before on a comparatively small scale. Thus the only goal attainable was the development of a uranium pile producing energy as a prime mover—in fact, future work was directed entirely towards this one aim."

Speer's decision put an end to the nightmare which had been afflicting Heisenberg and his collaborators. Till then they had never ceased to fear that another research team—for example Diebner's group, which worked in Thüringen—might be able to induce Hitler, after all, to embark on the construction of atom bombs. But probably those who were in favor of such a thing would in any case have had their plans wrecked by Hitler's shortsightedness. For that "leader of genius" had given the order in 1942, when he believed final victory to be imminent, that no armament projects were to be set on foot which could not guarantee the production of weapons ready for use in the field within six weeks. Weizsäcker refers in the following terms to the inadvertent but welcome support he and other passivists received from the head of the Army weapons department: "I remember that Schumann, a bad physicist but an extremely skillful tactician, once strongly advised us never to breathe a word about atom bombs to the high-ups if we could possibly avoid it. He said: 'If the Führer hears about it he'll ask "How much time will you need? Six months?" And then, if we haven't got the atom bomb in six months, all hell will break loose!' "

In 1942 Heisenberg and his friends won a domestic victory. In the small town of Seefeld in the Tyrol a dispute arose between the adherents of the "German Physics" promoted by the Nazis, and the modernists. It was compared by the participants with the religious disputations of past centuries. The contest ended with the triumph of the moderns, who had not, before then, been officially recognized by the Third Reich.

A handwritten summary of the arguments used was found by Goudsmit among Weizsäcker's other papers. The document, intended for the political authorities, represents a compromise to which even a great mind like Weizsäcker's did not scorn to resort. He began by using the language of official propaganda. He wrote: "The opinion was however expressed at the Seefeld meeting that the transference, mainly contrived by the Jewish Press propaganda of the Weimar period and the Jewish followers of Einstein, of the relativity theory in physics to problems concerned with views of the world in general, should be rejected. . . ." But after jotting down this sentence Weizsäcker must have felt repelled by it, for he crossed out the passages about "Jewish Press propaganda" and "the Jewish followers of

Einstein." Yet on reading the passage through again he apparently found the excisions too risky, as he placed a row of dots under one, at any rate, of the two adjectives, "Jewish," to show that it was to stand.

This reluctance by Weizsäcker to come down firmly on one side caused Goudsmit to doubt whether he had really been actively opposed to Hitler. In fact, even today Goudsmit is still convinced that Heisenberg and Weizsäcker would have constructed atom bombs if they had been in a position to do so. Other physicists, too, both in and out of Germany, have still not quite forgiven Weizsäcker for the diplomatic skill which he alleged he was obliged to practice in order to keep his real sentiment secret. Their resentment has led them, moreover, to forget what he, in his cautious manner, actually did in the struggle against Hitler.

The Alsos mission could not rest content with the discovery of Weizsäcker's papers. For it was thought in Washington that the papers might have been deliberately left behind, as a typical German strategem. It was considered that until all important physicists had been arrested and all laboratories occupied some doubt must still remain as to whether work might not still be going on somewhere in Germany on atom-bomb production. Goudsmit always insisted that no one but Heisenberg could be the brain of the German uranium project. The American military authorities dryly suggested that possibly other German physicists of whom Goudsmit had never heard might be secretly working on a weapon of this kind. But he repudiated their skepticism with the facetious retort: "A paper hanger may perhaps imagine that he has turned into a military genius overnight, and a traveler in champagne may be able to disguise himself as a diplomat. But laymen of that sort could never have acquired sufficient scientific knowledge, in so short a time, to be able to construct an atom bomb."

Consequently Heisenberg continued to be the most important "military objective" of his old friend Goudsmit. Wherever

Heisenberg might be, there too must be the main laboratory for the German atomic-armament plans. But where was he?

In the winter of 1943-44 Heisenberg and some of his collaborators had built a small model reactor in one of the cellars of his Dahlem Institute. The apparatus worked on one and a half tons of uranium and the same weight of "heavy water." The data furnished by this reactor had to be taken down, for the most part, during severe air raids. Since reliable work under these conditions proved practically impossible the whole Institute was gradually transferred to the little town of Hechingen, in a supposedly safe area, near the Swabian Alps, overlooked by the ancestral home of the Hohenzollerns. The tall boiler-house belonging to a Stuttgart brewery, where enormous beer vats had hitherto been stored, was papered from floor to ceiling with silver foil and provided with a high-voltage plant. Offices and workshops were established in the wing of a textile mill.

The next step was to find an even safer place for the construction of a new uranium burner. The Munich professor Walter Gerlach, though hostile to the regime, had taken over, shortly before the end of the war, control of the nuclear-physics division of the Research Council of the Reich. He remembered that as a lecturer at Tübingen he had known the picturesque little town of Haigerloch, situated on two precipitous heights over the River Eyach. Like many of his colleagues he had often paid a visit to it in lilac time. The innkeeper at the "Swan" made no difficulties about leasing to him a storeroom hollowed out in the rock of the steep hill on which the Castle stood. There, in February 1945, the construction of a new German pile was begun.

In all Germany there were few such operatically romantic sites as Haigerloch. The saga writer Gustav Schwab was of the opinion, after an excursion to this rugged scenery, that "this town is really absurd." The patriotic Pfeiffer composed the following couplet:

> He summons up the Fiend and cries,
> "Come, here let Haigerloch arise!"

At this spot, which had hardly changed since the Middle Ages, the most modern of German power stations was now built—an atom burner containing cubes of uranium and heavy water, with a graphite jacket. Every morning the physicists cycled in from Hechingen, some ten miles away, to their work in the electrically lit underground chamber in the rock. During the period of waiting for the moment at which the reactor would begin to send out rays and deliver power Heisenberg went up to the half-Gothic, half-baroque church attached to the Castle above and played Bach fugues on the organ.

"It was the most fantastic period of my life," one of those who took part in the experiments recalls. "I have never so often been obliged to think of Gounod's *Faust* and Weber's *Freischütz* as I was in that extravagantly romantic setting." But the tests carried out led to only partial results. For there was not enough uranium for the point to be reached at which the chain reaction builds up "criticality." The consignments of uranium cubes from Berlin and from Ilm in Thüringen—where another uranium burner had been constructed under Diebner's directions—could not get through, by that time, as far as Haigerloch.

As soon as the Alsos organization learned where Heisenberg had taken refuge Colonel Pash proposed that a parachute detachment should precede the Allied armies to Hechingen and Haigerloch, drop on the atomic scientists at work there and "make sure" of them and their notes. But Goudsmit had meanwhile come to the conclusion, as a result of the papers that had been discovered and the examination of witnesses, that such a step was quite unnecessary. As he truly prophesied, "There is so little danger in what is going on at Hechingen and Haigerloch that I consider the German project not worth even the sprained ankle of a single Allied soldier."

After the front had collapsed Pash was less afraid of the Germans than of the French, whose occupation zone happened to include Hechingen. He was determined to steal a march on them. Hastily assembling a small unit of shock troops, together with two tanks, a few jeeps and some heavy transport, he "captured" Hechingen on April 22, 1945, at half-past eight in the morning, eighteen hours before General de Lattre's troops marched in. Pash's T (technical) Group occupied Haigerloch the same day. There a last-minute attempt had been made to get the uranium cubes away into safety by putting them on an ox-cart and hiding them in a barn under the hay. But one of the German atomic scientists—actually a man who had always been fond of seasoning his conversation with patriotic maxims —was anxious to gain the good graces of the new masters without delay. Consequently he revealed the hiding place. Some of the uranium had, as a matter of fact, been previously spirited away by young Hechingen peasants, who subsequently tried to sell it to the French occupation authorities. But they were arrested by the French and given heavy sentences for theft. A few days later the remains of the uranium burner in the rock chamber were blown up by an Allied detachment in the absence of such instructions from Goudsmit and much to his annoyance.

In the course of the activities of the Alsos mission the group captured eight members of the two Kaiser Wilhelm Institutes, for physics and chemistry respectively. They included Otto Hahn, the discoverer of nuclear fission, the Nobel prize winner Max von Laue and also C. F. von Weizsäcker. But Heisenberg could not be found. He had mounted a bicycle at three in the morning and made off in the direction of Upper Bavaria, where his family was living. He wanted to be with them at that period of the last of the fighting. On the way he was nearly arrested by a fanatical SS man. Fortunately the latter allowed himself to be bribed with a package of Pall Mall cigarettes which in some roundabout fashion had found its way into Heisenberg's possession from the supplies belonging to Marshal Pétain, who had been interned in the castle of Sigmaringen, not far from Hechingen.

Colonel Pash and Goudsmit had again failed to run down the prize quarry they were after. By way of consolation, however, they discovered in the office of the "one that got away" a photograph of Heisenberg cordially shaking hands with none other than Goudsmit himself. The date was 1939, the occasion the last visit of Heisenberg to America and the place the home of the present chief of the Alsos mission at Ann Arbor.

Another senior officer of the Army Intelligence Service, General Harrison, had taken part in the capture of Hechingen. His emotions, when he examined this "remarkable photograph," were very mixed. Goudsmit reports: "The Colonel and the General entered Heisenberg's office. He was not there. But the first thing they saw, to the consternation of the General, was a photo of Heisenberg and myself standing side by side. . . . Egged on by Colonel Pash the General almost began to believe that I could not be trusted and that I had close contact with the enemy. I could have helped him out, I suppose, but that didn't seem quite the moment to explain about the international 'lodge' of the physicists."

Eleven

Atomic Scientist versus Atomic Bomb (1944-1945)

Shortly after Goudsmit had discovered Weizsäcker's papers dealing with the German atomic project he went for a walk with a major who had been attached to the Alsos group in the capacity of liaison officer with the War Department.

"Isn't it wonderful," Goudsmit remarked, "that the Germans have no atom bomb? Now we won't have to use ours."

The professional soldier's retort shocked Goudsmit, for, out of his many years' experience of the military mind, he prophesied: "Of course you understand, Sam, that if we have such a weapon we are going to use it."

Thoughts similar to Goudsmit's troubled the atomic scientists who read at General Groves's head office, in their capacity as experts, Goudsmit's detailed reports from the seat of war. Each report of the Alsos agents, who by this time had captured in Heidelberg, Celle, Hamburg and the city of Ilm in Thüringen all the members of the Uranium Society, including finally even Heisenberg himself, found at his house near Urfeld, made it clear that the Germans really did not possess any atom bombs. They had not even created the preliminary practical conditions for the construction of such a weapon. The Alsos mission's reports of the non-existence of a German atom bomb were naturally "top secret." But no security measure, however strict, could stop this amazing news from going the rounds of all the Allied laboratories, where it was eagerly discussed.

The reports confronted the atomic scientists, intellectually and psychologically, with an entirely new situation. The assumptions on which they had started work were no longer valid. Could any further work on the bomb be politically and morally justified at all now? Of course not! For the Japanese, who were at present the only serious adversaries the Allied Nations had left, were in no position, it was known for certain, to develop any such weapon.

On the other hand it would have been contrary to the spirit of modern science and technology to refrain voluntarily from the further development of a new field of research, however dangerous it might be for the future, and leave it only half explored. New grounds had therefore to be provided for the political and moral justification of the continuance of work, even under these changed conditions, in the atomic-research laboratories. Such arguments were soon forthcoming. They ran somewhat as follows.

"If we don't now develop this weapon and demonstrate to the world, by public experiment, its appalling nature, sooner or later some other unscrupulous power will attempt, unobtrusively and in all secrecy, to manufacture it. It will be better for the future peace of the world if humanity at least knows where it stands." Such was the attitude, for instance, of Niels Bohr in the confidential discussions which arose. But an even stronger argument for justifying further research went as follows: "Humanity needs the new source of power which we have discovered and developed. All we have to do is to take care that in future it shall be used for peaceful purposes instead of for destruction."

Such problems were most intensively debated in the Metallurgical Laboratory of the University of Chicago. Since, after 1944, the main task of development had been transferred to Oak Ridge, Hanford and Los Alamos, time could be given in Chicago, where the atomic project had registered its first important results, to the consideration of the actual consequences to be expected from the new invention. It was among the Chicago scientists, too, that the first protests were heard later against the proposal to use the bomb in the war against Japan. They were also the first to consider thoroughly the possibilities of the international control and peaceful development of atomic power.

It was in Chicago, as early as the summer of 1944, that a committee of atomic scientists was formed under the chairmanship of Zay Jeffries, who in peacetime had been one of the directors of General Electric. The committee composed a number of reports on the potentialities and perils of the new, epoch-making discoveries. These reports were submitted to General Groves on December 28, 1944, under the title of "Nucleonic Prospects."

Independently of this committee Bohr had, since the beginning of 1944, been studying the political problems involved in the discovery of the "new power." The great Danish scholar did not share the current optimism with regard to the future relations between the chief partners in the alliance opposed to the Axis. He foresaw that there might be friction and conflict between East and West after the war. Agreement among the three great powers, the United States, Britain and Soviet Russia, as to common control of all applications of atomic energy seemed to Bohr more easily attainable before completion of the atom bomb or its actual employment in war.

On August 26, 1944, at four o'clock in the afternoon, Bohr was received by Franklin D. Roosevelt at the White House. His object was to discuss these questions, likely to be matters of life and death in the very near future, on the basis of a detailed memorandum which he had addressed to President Roosevelt and to Churchill on July 3.* The scientist intended to point out that the new power might make an important contribution to reconciliation of the mutually opposed standpoints of Bolshevist Russia and its partners, whose political and economic organizations were so different. He had suggested that, in order to initiate preliminary and for the time being unofficial contacts, the international relations between sci-

* For the text of this memorandum see Appendix A, p. 344.

entists, which had been only temporarily interrupted, should now be utilized. Bohr hoped for the rise of a family of nations through the spirit of a reunited family of atomic scientists.

The course that this conversation took is unknown. For Roosevelt made it a rule never to record such private interviews. And Bohr himself still feels, even today, that it is his duty to keep the secret. It is clear, nevertheless, that the President did not agree to Bohr's suggestion. He may have considered it premature. Or it may simply have been the case that Bohr, who could be extraordinarily persuasive in long conversations, did not manage to convey his meaning definitely enough at a brief meeting.

The latter hypothesis seems plausible on the evidence of the account given by the physicist Lord Cherwell, Churchill's scientific adviser, of a similar interview granted by the British Prime Minister to Bohr. He recalled that Churchill listened to the scientist for half an hour in silence. But at the end of that time he suddenly stood up and broke off the audience, before Bohr had finished his low-toned and circumstantially detailed explanations. The Prime Minister is then said to have turned to Cherwell and asked, with a shake of the head: "What is he really talking about? Politics or physics?"*

Alexander Sachs felt himself, like Bohr, that he shared responsibility for the development of the atom bomb. After Bohr had taken the above-mentioned action Sachs tried to persuade the American President to commit himself as soon as possible to a definite attitude with regard to the new weapon. Roosevelt's "anonymous adviser" had talked the President, five years previously, into giving the signal for the start of its construction. He now drew up a memorandum, which he read out to Roosevelt

* It is not generally known that the Pope uttered a warning against the destructive use of atomic energy in a speech to the Pontifical Academy of Science on February 21, 1943, a date at which the ordinary public still knew nothing of the atom bomb. In April 1946 Monsignor Sheen made the following comment on that speech: "It is to be noted that the Holy Father not only knew about atomic energy and something of its power, but he also, exercising his office as Chief Shepherd of the Church, asked the nations of the world never to use it destructively. This counsel was not taken. This moral voice was unheeded."

in December 1944, dealing with the first use to be made of it. After a long discussion the two men agreed on the following conclusions, here given in the version submitted by Sachs to Robert P. Patterson about a year afterwards.

> Following a successful test, there should be arranged (*a*) a rehearsal demonstration before a body including internationally recognized scientists from all Allied countries and, in addition, neutral countries, supplemented by representatives of the major [religious] faiths; (*b*) that a report on the nature and the portent of the atomic weapon be prepared by the scientists and other representative figures; (*c*) that thereafter a warning be issued by the United States and its allies in the Project to our major enemies in the war, Germany and Japan, that atomic bombing would be applied to a selected area within a designated time limit for the evacuation of human and animal life; and finally (*d*) in the wake of such realization of the efficacy of atomic bombing an ultimatum demand for immediate surrender by the enemies be issued, in the certainty that failure to comply would subject their countries and people to atomic annihilation.

Did Alexander Sachs only imagine that Roosevelt had agreed to this plan of December 1944? At any rate the President never seems to have mentioned to Stimson, his Secretary of War at the time, the existence of this draft of a directive to govern the use of the atomic bomb, though the problems concerned were exhaustively discussed between Roosevelt and Stimson later on. Stimson, one of the few politicians who knew anything about the work being carried out on the bomb, saw the President for the last time on March 15, 1945. On that occasion the talk was chiefly of "X," as Stimson sometimes designated the new weapon for security reasons in written documents. He noted in his diary: "I went over with him the two schools of thought that exist in respect to the future control after the war of this project, in case it is successful, one of them being the secret close-in attempted control of the project by those who control it now, and the other being international control based upon freedom of science. I told him that those things must be settled before the projectile is used and that he must be ready with a statement to come out to the people on it just as soon as that is done. He agreed to that."

General Groves, for his part, had no doubt whatever that the bomb would be used for war purposes as soon as it was ready. At the beginning of 1945, when several atom bombs were expected to become available within a few months, the head of the Manhattan Project reported to his immediate superior, General George Marshall, Chief of Staff. Groves proposed that detailed plans for the employment of the bomb in war should now be drawn up and suitable senior officers entrusted with the direction of these preliminary studies. But Marshall had been so satisfied with the services so far rendered by Groves that he replied, so Groves relates: "Can't you see to all that yourself?"

This reply was more than a question. It was an order. "GG" was only too glad to obey it. He had long since outgrown his status as a military manager specializing in building construction. He now considered himself to be a practical nuclear physicist, capable of making scientific decisions, as well as a diplomat (working, for instance, against the policy planned by his government of collaborating with the British on atomic problems). He intended henceforth to be also a strategist and—since employment of the atom bomb in warfare would raise questions of the highest political importance—a statesman into the bargain.

Certainly up until now his performance had seemed gigantic. Under his direction factories had arisen at Oak Ridge that were longer than those to be found anywhere else in the United States. Works inspectors had to use bicycles to patrol them. At Hanford 60,000 laborers had built one of the largest chemical works in the country. At Los Alamos seven divisions were employed on the mysterious "end-product."* Literally thousands of new in-

* In the spring of 1945 the following divisions were active at Los Alamos: Theoretical Physics Division (Director: H. Bethe), Experimental Nuclear Physics Division (Directors: J. W. Kennedy and C. S. Smith), Ordnance Division (Director: Captain W. S. Parsons), Explosives Division (Director: G. B. Kistiakovsky), Bomb Physics Division (Director: R. F. Bacher), Advanced Development Division (Director:

ventions and patents had been developed in the course of the work. The description alone of the most important new processes developed at Hanford would have filled thirty stout volumes. Was the practical application of the result of years of strenuous efforts by 150,000 people, the introduction of a weapon that had involved the expenditure of two billion dollars, now to be voluntarily renounced? General Groves did not even trouble to discuss the idea. In his view it was too silly to consider. An atomic scientist who was working in close contact with him at that time states that from 1945 on Groves gave the impression of being obsessed by one intense fear, that the war would be finished before his bomb could be. Accordingly, even after the capitulation of Germany he continued to exhort his collaborators with the incessant slogan: "We must not lose a single day."

As early as the spring of 1945 a study group within the Manhattan Project was given the task of selecting a target for the first employment of the bomb. The group comprised mathematicians, theoretical physicists, specialists on the effects of explosions and meteorological experts. This team, mainly composed of scientists, including Robert Oppenheimer, came to the conclusion, according to a report published later in a limited edition by the Manhattan Engineer District, that targets for this peculiar type of bomb must satisfy the following conditions:

(*a*) Since the atomic bomb is expected to produce its greatest amount of damage by primary blast effect and next greatest by fires, the targets should contain a large percentage of closely built frame buildings and other construction that would be most susceptible to damage by blast and fire;

(*b*) the maximum blast effect of the bomb is calculated to extend over an area of approximately one mile in radius. Therefore the selected targets should contain a densely built-up area of at least this size;

(*c*) the targets selected should possess high military and strategic value;

Enrico Fermi) and the Chemistry and Metallurgy Division. Each division was sub-divided into groups under their own leaders.

(*d*) the first target should if possible be one that has escaped earlier bombardments, so that the effect of a single atomic bomb can be ascertained.

It was further resolved that four Japanese cities should be deliberately spared bombardment by the American formations which by 1945 could reconnoiter any target they pleased in Japan with practically no resistance. This deceptive period of grace was granted these four cities so they could be doomed to a ruin all the more dreadful by the new bomb.

On the short list of targets for the atom bomb, in addition to Hiroshima, Kokura and Nigata, was the Japanese city of temples, Kyoto. When the expert on Japan, Professor Edwin O. Reischauer, heard this terrible news, he rushed into the office of his chief, Major Alfred MacCormack, in a department of the Army Intelligence Service. The shock caused him to burst into tears. MacCormack, a cultivated and humane New York lawyer, thereupon managed to persuade Secretary of War Stimson to reprieve Kyoto and have it crossed off the black list.

By the spring of 1945 pilots at Wendover airfield in Utah were being trained for the first raid with atomic bombs. At the same time Leo Szilard, whose initiative had set in motion the construction of this weapon, made a last attempt to recapture and seal up again in its bottle, before it could do any harm, the sinister "djinn" which he himself, like the fisherman in the *Arabian Nights,* had released. Later he gave a remarkably frank account of his feelings in those months:

"During 1943 and part of 1944 our greatest worry was the possibility that Germany would perfect an atomic bomb before the invasion of Europe. . . . In 1945, when we ceased worrying about what the Germans might do to us, we began to worry about what the government of the United States might do to other countries."

There had been a complete reversal of the situation. In the summer of 1939 Szilard had visited Einstein in order to obtain his assistance in persuading the government to construct an atom bomb as a preventative measure. Now, more than five years later, he again approached Einstein, this time to explain to him the utterly different global situation and to indicate in broad outline—for to give details would have been to infringe security regulations—the possibility that the United States might initiate an atomic-armaments race. Einstein proceeded once more to sign a warning letter intended, together with a circumstantial memorandum by Szilard, for President Roosevelt. Szilard had written his opinion that any momentary military advantage the bomb might bring to the United States would be offset by grave political and strategic disadvantages. But neither Einstein's final letter nor Szilard's eloquent warning against employment of the bomb ever came to the notice of the President. Both documents were still lying on his desk, untouched, when suddenly, on April 12, 1945, he died.

Szilard could at first see no opportunity whatever of laying his request, urgent even in these circumstances and especially so in view of the advancing preparations for atomic bombardment, before the new President, Harry S. Truman. It was said in Washington that in those first weeks the former senator would at most be accessible to the somewhat restricted circle of residents in his own state, Missouri. Luckily one of Szilard's scientific collaborators at Chicago came from Kansas City, Missouri. This man knew Truman's secretary, Matt Connelly, who was, of course, a citizen of Missouri, and took Szilard to see him.

The new President had only recently himself been fully informed by Stimson, on April 25, of the strictly secret bombing project. Overwhelmed as he was by a mass of novel and unexpected tasks, Truman, understandably enough, had no time just then for a personal interview with Szilard. Connelly therefore sent the scientist to Justice James F. Byrnes, an influential Democrat, though at the moment he held no official appointment.

Byrnes had practically no experience in international politics at that time, though he was appointed Secretary of State by President Truman later that summer. However, Byrnes had considerable influence in domestic affairs. A former justice of the

Supreme Court, he had headed the Office of War Mobilization in 1944.

Szilard accordingly had to set out, with his memorandum and a copy of Einstein's letter, to the southern city of Spartanburg in South Carolina, "Jimmy" Byrnes's political stronghold. There the atomic scientist, whose fate it had been to become a citizen of the world and who bore a burden of anxiety for all mankind, confronted the successful politician, who knew that he himself was about to enter upon a great advance in his career. The considerations in Szilard's memorandum related to a future probably much further ahead than the other cared to look. They involved such surprising and unheard-of measures as a partial surrender of national sovereignty, with Soviet police on American soil and American police on Soviet soil, international action to supervise prospecting for uranium and production of atomic power. In the sleepy atmosphere of Spartanburg such ideas sounded like unpractical or even hysterical fancies.

Szilard soon saw that Byrnes had no sympathy with his arguments, though he concealed his lack of interest in them under the cover of conventional amiability which a professional politician always has at call. "Are you not worrying too much and quite unnecessarily about all this?" he asked his guest, whose foreign name he found so hard to pronounce. "So far as I know, there is not even any uranium to be found in Russia."*

Only a few weeks later President Truman appointed James Byrnes to the office of Secretary of State.

Stimson, the Secretary of War, had asked President Roosevelt, shortly before the latter's death, for a directive dealing with the employment of the first atom bombs and the future of the project for their manufacture. But he had never received it. Consequently, at his first interview with President Truman at the end of April 1945, Stimson strongly urged the setting up

* Harrison Brown comments on Szilard's visit to Byrnes: "If he had been taken as seriously by Mr. Brynes as he had been taken seriously by Mr. Roosevelt more than five years previously the whole course of history might have been altered."

at the earliest possible moment of a committee of experts to advise the President on these questions. The news, when it became known in the laboratories of the Manhattan Project, was a great relief to all the scientists who dreaded that the bomb might be used and were in favor of early international control. But their disappointment was equally great when they learned who were to serve on the committee. Five of the members were prominent politicians. They comprised the Secretary of War, Henry L. Stimson, his deputy, George L. Harrison, James Byrnes, as Truman's personal representative, Ralph A. Bard, representing the Navy, and William L. Clayton for the State Department. In addition, there were three scientists who, since 1940, had been in charge of the entire organization devoted to research for military purposes. Their names were Vannevar Bush, Karl T. Compton and James B. Conant. This main committee was reinforced by a special committee of atomic experts, the "scientific panel": J. Robert Oppenheimer, Enrico Fermi, Arthur H. Compton and Ernest O. Lawrence. The seven scientists, with the possible exception of Fermi, were known among their colleagues for their tendency to play ball with politicians and the military. They were not trusted to represent fairly the views of many, if not actually the majority, of those at work on the Manhattan Project. A suggestion that the Nobel prize winner Harold C. Urey, who possessed the confidence of a great many atomic scientists, especially of the younger generation, should be included in the scientific panel was not adopted.

Such then was the body, given the intentionally vague name of the "Interim Committee," which met on May 31 and June 1 to consider, in the words of its orders from Marshall, "atomic energy not simply in terms of military weapons but also in terms of a new relationship of man and the universe." Arthur H. Compton remembers that the scientific panel, to which he himself belonged, was not called upon to decide the question *whether* the new bomb should be used but only *how* it should be used. At this first consultation the four atomic experts unfortunately kept strictly to their limited instructions, instead of suggesting, on their own account or as the mouthpieces of many of their professional colleagues, that the bomb should not be used in warfare. Compton reports, with regard to Oppenheimer's stand before the Interim Committee: "He was giving a technical reply to a technical question." The reply referred to estimated that some 20,000 people would be killed the first time the bomb was dropped. The fact that employment of the bomb, in Compton's words, "throughout the morning . . . seemed to be a foregone conclusion" was due, above all, to the influence of a man whose name does not appear in the list of members and who is also not mentioned by Stimson in his subsequent account of the proceedings. This man's name was Leslie R. Groves.

"It would not have looked well," Groves explains, "if I had been officially appointed to serve on a committee of civilians. But I was present at all its meetings and I always considered it my duty to recommend that the bomb should be dropped. After all, great numbers of our boys were dying every day at that time in the war against the Japanese. So far as I am aware, none of the scientists who opposed the dropping of the bomb had any near relatives in the field. So they could very well afford to be soft." *

The outcome of the committee's deliberations was in any case a complete victory for Groves. On the conclusion of the meetings the following recommendations were forwarded to President Truman:

(1) The bomb should be used against Japan as soon as possible.

(2) It should be used on a dual target—that is, a military

* Compton states that General Marshall did in fact make the point that it would be better for the postwar safety of the nation if the existence of the atomic weapon could be kept secret. But he did not press the point after being assured by the scientists that no such secrecy could be maintained for long. At lunch after the session Compton raised the question whether a non-military demonstration of the weapon could not be arranged. The majority of those present at the meal rejected the idea for one reason or another. He then, with surprising alacrity, withdrew his proposal.

installation or war plant surrounded by or adjacent to houses and other buildings most susceptible to damage.

(3) It should be used without prior warning of the nature of the weapon.

The third recommendation was considered by Ralph A. Bard, who represented the Navy at these discussions, to be so unfair that he afterwards withdrew his agreement to the point in question. His was the only voice that registered even a mild protest.

It was probably no accident that the representative of the American Navy proved more accessible to humane considerations than his colleagues. The United States Navy has always shown utmost respect for the rules of chivalrous combat. A typical case of this attitude occurred when the Judge Advocate General of the Navy Department was asked, shortly before the decision to drop the atomic bomb, to give a juridical opinion as to the use of a "biological weapon" recently developed in a naval research laboratory. The product concerned was a biochemical preparation which would have destroyed the whole of Japan's rice harvest and would thus have brought the country near to starvation if it had been applied from the air. The jurist's opinion was that so inhuman a weapon would be an offense against the laws of war and consequently unworthy of the Navy.

But no attempt was ever made to obtain any such legal pronouncement from any competent source before the atom bomb was employed.

The recommendations made by the Interim Committee were, of course, strictly secret. They filtered through, however, to Chicago, Oak Ridge and Los Alamos, where the younger scientists, in particular, continued to speak more and more openly against the use of the bomb. At Chicago the University appointed a committee to discuss and report in detail upon the "social and political consequences of atomic energy." Its chairman was the Nobel prize winner James Franck, the former

Göttingen professor. Apart from Franck, it was probably Szilard and the biochemist Eugene Rabinowitch who contributed most of the ideas in the memorandum, which afterwards became known as the "Franck Report." *

"It was unbearably hot in Chicago at that time," Rabinowitch recalls. "As I walked through the streets of the city, I was overcome by a vision of crashing skyscrapers under a flaming sky. Something had to be done to warn humanity. Whether on account of the heat or my own inward excitement, I could not sleep that night. I began writing our report long before daybreak. James Franck had given me a draft of one and a half pages as his contribution. But my own treatment of the matter became very much more detailed."

The seven scientists from Chicago began their report,† forwarded to the Secretary of War in the form of a solemn petition on June 11, 1945, by declaring that they did not presume to speak authoritatively on problems of national and international policy. They proceeded to affirm that only in the capacity of a small group of citizens cognizant of a grave danger of which the rest of mankind was unaware had they resolved to act. They were the more impelled to do so, they continued, because as modern men of science, unlike the investigators of former ages, they were unable to hold out the prospect of any effective defense against the new weapons, so far did their destructive power exceed anything hitherto known. Such a defense could not on this occasion, it was added, be provided by scientific inventions, but only by a new political organization of the world.

This passage was followed by an amazingly accurate forecast

* Compton states that "within the Committee as a whole there had been sharp divisions as to whether the bomb should be used in the Japanese war. The draft as prepared, however, included only the argument against the bomb's military use. It was submitted, therefore, not as a report of the Committee, but as representing the view of the Committee's chairman and of certain others who shared it with him." The seven men concerned were James Franck, D. Hughes, L. Szilard, physicists, T. Hogness, E. Rabinowitch, G. Seaborg, chemists, and a biologist, C. J. Nickson.

† For complete text see Appendix B, p. 348.

of the armaments race to be expected—as was later proved. In order to avoid such a contingency, the report continued, efforts should immediately be made to establish control of armaments on a basis of mutual trust. It was just this essential confidence which would be destroyed at the start if the United States were to make a surprise attack on Japan with a bomb which would be certain, like the German rocket missiles, to slaughter soldiers and civilians without distinction. The seven scientists warned the Secretary: "Thus the military advantages and the saving of American lives achieved by the sudden use of atomic bombs against Japan may be outweighed by the ensuing loss of confidence and by a wave of horror and repulsion sweeping over the rest of the world and perhaps even dividing public opinion at home."

The Franck Report proposed that, instead of the atomic bombardment of Japan as planned, a demonstration of the new weapon might best be made, before the eyes of representatives of all the United Nations, in a desert or on a barren island. The report continued: "The best possible atmosphere for the achievement of an international agreement could be achieved if America could say to the world: 'You see what sort of a weapon we had but did not use. We are ready to renounce its use in the future if other nations join us in this renunciation and agree to the establishment of an efficient international control.'"

Compton believes, to the best of his recollection, that Franck himself came to Washington and handed him the report, which Compton then immediately passed on to George L. Harrison, Stimson's deputy. The urgency of this new appeal and the high respect in which the seven members of the Franck committee were held caused Stimson at once to refer the document to the scientific panel of atomic experts which had been consulted a fortnight before.

At that moment the four men concerned—Compton, Fermi, Oppenheimer and Lawrence—undoubtedly had it in their power, by agreeing to the suggestion of their colleagues at Chicago, at least to query and perhaps even to prevent the atomic bombardment of buildings and human beings occupying a military target in Japan. The panel met, for the second time, at Los Alamos on June 16, 1945. Oppenheimer subsequently gave the following account of its deliberations:

> The other two assignments which the panel had—one was quite slight. We were asked to comment on whether the bomb should be used. I think the reason we were asked for that comment was because a petition had been sent in from a very distinguished and thoughtful group of scientists: "No, it should not be used." It would be better for everything that they should not. We didn't know beans about the military situation in Japan. We didn't know whether they could be caused to surrender by other means or whether the invasion was really inevitable. But in back of our minds was the notion that the invasion was inevitable because we had been told that. I have not been able to review this document, but what it said I think is characteristic of how technical people should answer questions.
>
> We said that we didn't think that being scientists especially qualified us as to how to answer this question of how the bombs should be used or not; opinion was divided among us as it would be among other people if they knew about it.* We thought the two overriding considerations were the saving of lives in the war and the effect of our actions on the stability, on our strength and the stability of the postwar world. We did say that we did not think that exploding one of these things as a firecracker over a desert was likely to be very impressive.

The initiative taken by the seven men of Chicago was thus repudiated and the hope of stopping the atomic bombardment of Japan dwindled away to almost nothing.

Such was the background, one of petitions that failed and of growing disappointment among Allied atomic scientists, against which it is necessary to judge the thoughts and actions of a certain theoretical physicist, then thirty-four years old. Four and a half years later, at the end of January 1950, he was to

* It was E. O. Lawrence who objected most strongly to the dropping of the bomb. He did so, according to Compton, because some of his pupils had been Japanese.

become notorious as the central figure in a great espionage scandal. This man was Klaus Fuchs.

He was the son of a German Protestant pastor, who was a member of the Quaker community and a Christian Socialist. As one of the most important of the team of British atomic scientists Fuchs had arrived in the United States at the end of 1943 and at Los Alamos in December 1944. While a pupil of Max Born in Edinburgh and as assistant to Rudolf Peierls in Birmingham, Fuchs had been regarded as a somewhat taciturn and unsociable character. But in the restricted Los Alamos circle of scientists of all nations he had blossomed forth as never before. He made many friends and rendered personal services to his colleagues in all sorts of discreet ways. He was always ready to help out as a baby sitter, to execute shopping commissions in Santa Fe and in every possible manner to comply with the Christian precept to love one's neighbor. "He was one of the kindest and best-natured men I ever met," declares Dorothy McKibben to this day.

His colleagues remember how intently Fuchs listened to the discussions, more and more frequent after the beginning of 1945, on the political and social consequences of the bomb. He seldom made any contribution to them. He only interrupted on one occasion. When someone said: "We ought all to refuse to go on working, because the government will be breaking its unwritten contract with us if it uses the bomb for the purposes of military aggression," Fuchs retorted dryly, with devastating objectivity: "It's probably too late for that now. The whole thing's already in the hands of the technicians."

He may have been impelled less by an arrogant self-sufficiency than by sympathy with his disappointed colleagues when in February and June 1945, at Cambridge and Santa Fe, he passed on everything he knew about the atom bomb to the Soviet agent Raymond, alias Harry Gold. Perhaps he thought: "Others talk, hope, wait and are repeatedly disappointed, because they don't understand the true nature of political power. Well, I'm going to act Maybe I could prevent another war."

At a later date, after his arrest, he declared that he was sorry to have played a double game with his friends. If such scruples had occurred to him as early as the spring of 1945, he must have consoled himself with the reflection that in those days all these friends had repeatedly and emphatically stated, during their debates, that the new weapon forced mankind to transcend national boundaries in their thinking and to take altogether exceptional action, contrary to hitherto-accepted ideas of patriotism and national loyalty.

No doubt the other atomic scientists intended later on, with the permission of the state and public opinion, to commit this treason to conceptions which they considered historically out of date. But such a plan must have appeared to Fuchs doomed to failure. What did he really think at that time? His own statement, made after his arrest, that he had acted schizophrenically, does not carry much conviction. He probably hoped by this declaration to escape the death sentence he must have feared and to be given a milder punishment as being "temporarily irresponsible for his actions."

We may be able to obtain a more accurate notion of what was actually in Fuchs's mind when he resolved to commit high treason if we consider the comments of his father and close friends, with whom he never lost touch, either before or after his arrest. Pastor Emil Fuchs, who had a private interview with his son after the war, in England, attempts today to explain his son's conduct as follows:

As a father I can understand his extreme inward distress at the moment when he realized that he was working for the manufacture of the bomb. He said to himself, "If I don't take this step, the imminent danger to humanity will never cease." He thus found a way out of a situation that seemed hopeless. Neither he nor I have ever blamed the British people for his sentence. He endures his fate bravely, with determination and a clear conscience. He was justly condemned under British law. But there must of course always be people from time to time who deliberately assume such guilt as his. The Prussian General Yorck did so in 1812 when he neutralized the Prussian Army by the Convention of Tauroggen. They have to bear the consequences of their resolute affirmation that they see a position more clearly than do those in whose power, at that juncture, the decision rests to deal with it. Should it not be clear by this time that my son acted with more accurate foresight in the interests of the British people than did their government? His action imperilled the highly paid and distinguished post he held and a still more distinguished career in the future. I can only have the greatest respect for the decision he took. Which of us can be certain how we would have decided in a similar situation? *

Margaret Hager, a close friend of the Fuchs family, wrote a long study of Klaus Fuchs, which she sent to certain acquaintances. It includes the following revealing passages:

As soon as one begins to examine this case it becomes clear that the moral ideas which we more or less intuitively apply, in a deplorably mechanical fashion, are not immediately adequate and effective in matters in which one does not take a one-sided view of loyalty but regards it as obligatory in a comprehensive sense. Loyalty is then not something to be shown to an individual against others but simply and in principle loyalty to mankind as a whole. The question is accordingly not, How could Dr. Fuchs act in such a way? but, How could a society and thus also—of necessity—a man act when confronted by circumstances, against his will and without his desiring it, with the problem of the best use of power here and now in this world? Fuchs was forced to become a traitor in these conditions.

But how would it be if he did the opposite to what he apparently did? He would then be guilty in the other direction. The atomic secret would be kept and the man would remain true to his oath. War, from a human standpoint, would be inevitable. Dr. Fuchs would then have taken the easy path. No one, not even the strictest

* Pastor Fuchs, however, did not discuss with his son, on this occasion, the latter's political activities. On this subject he wrote as follows to the author: "I certainly realized to the full how much distress his work on atomic research caused him. But in 1933 we had arranged once for all that we would always tell each other anything that ought to be known about illegal activities by anyone, though never more than what was necessary. Which of us can be sure that he can keep another's secret? Consequently, on this occasion too, we did not mention such things, though I guessed that something of the sort was worrying him and wondered how he would ever be able to reconcile himself inwardly with his whole situation."

moralist, would probably have been able to reproach him in that case. For no one would be aware that for one fateful moment the life and death of nations lay—humanly speaking—at his mercy. Subjectively, Fuchs must be judged wholly innocent. But . . . in the end the British people would of course also have to suffer the horrors of a third world war . . . and though he himself would have remained "innocent," it would have been at the expense of the nations of the earth.

But this argument does not prove that Dr. Fuchs was unjustly condemned. He was guilty according to contemporary moral ideas of loyalty. But he became so fundamentally, in order that nations, individuals and humanity at large might learn, in principle, from his case, where our present social organisation is taking us. . . . The circumstance does not turn Dr. Fuchs into a hero. But however little of a criminal or martyr he may be, he is clearly—let us understand the issue at its deepest level—a stopgap, safeguarding the inward development and transformation of the human race in its progress towards a more genuine, more truly creative humanity. . . . Apparent disloyalty can be deeper loyalty than the common type.

Fuchs himself made the following revealing comment after his arrest: "It appeared to me at the time that I had become a free man because I had managed to establish myself in an area of my being as completely independent of the surrounding forces of society."

It is obvious that he instantly lost this freedom again when he handed on his information to the agents of another power. He had only exchanged one loss of liberty for another.

Twelve

For They Know Not What They Do (1945)

Never was the pace at Los Alamos fiercer than after the capitulation of the Third Reich. "Our husbands worked almost incessantly," Eleanor Jette, the wife of a leading atomic scientist, remembers. She was a sort of local celebrity on account of her authorship of the sketches regularly performed on New Year's Eve in which the difficulties of life on the mesa and the peculiarities of certain of its prominent inhabitants were good-naturedly satirized.

But in June and July 1945 even she lost her sense of humor. It seemed the weather had conspired to frustrate the bomb builders. For weeks not a drop of rain fell. A dry, hot wind blew from the desert over the settlement. The grass withered. Foliage and pine needles dried up on the trees. Now and then the sky darkened and lightning flickered in the distance over the Sangre de Cristo Mountains. But the clouds did not open. Forest fires started on a number of occasions close to the town of laboratories. It was feared that the wind would carry sparks to the residential quarters, office buildings and workshops. They were still all timber constructed and would easily catch fire. But if this happened the only water available for extinguishing purposes was Ashley Pond, a small pool in the center of the settlement. The supply system did not now even suffice for the most necessary personal ablutions with this precious liquid. "We brushed our teeth with Coca-Cola," says a hospital nurse. "To cap it all some cases of chicken-pox chose this particular moment to break out among the schoolchildren. It became more important than ever for all the children to be able to wash as often as possible and that was just when there was a shortage of water."

A directive from Groves required the first bomb to be ready for testing by the middle of July and the second to be available for war purposes in August. Philip Morrison states: "I can personally testify that a date in the neighborhood of the 10th of August was the mysterious deadline which we, who were working daily on the job of finishing the bomb, had to observe at all costs, irrespective of money troubles or the mass of development work still to come."

The rush of business, the heat and the scarcity of water combined to make everyone irritable. Mrs. Jette reports: "One day, as I was passing an old acquaintance, I quite innocently said 'Good morning!' He instantly turned upon me in a fury, exclaiming: 'What's so *good* about the morning?' "

During this final stage of the construction of the atom bomb two young physicists, both of them, incidentally, in the same age-group as Klaus Fuchs, became particularly conspicuous. They were the exceptionally tall Californian Luis W. Alvarez and the skinny Louis Slotin, born in Canada of Russian parents who had fled there from the pogroms. Both these scientists were "war babies" who had first come to be experts at their trade by doing war work and achieved their first big results in arms laboratories. In their eyes the "new power" was neither so wonderful nor so terrible as it seemed to the veterans with whom they worked. They had not, therefore, much sympathy with the doubts that afflicted their seniors in these last few months.

Alvarez, the son of a well-known surgeon at the Mayo clinic, had come to Los Alamos rather late, after distinguishing himself in the secret radar laboratory attached to the Massachusetts Institute of Technology. He had made some important discoveries there, including the invention of a bomb-aiming device and the ground-control approach system now in use at practically all airfields. On the Hill at Los Alamos he and his still younger research team had succeeded in constructing the complex release mechanism of the bomb, accurate to one-millionth of a second.

The testing of this apparatus, so far as it could be done in the laboratory, was regarded as one of the most dangerous jobs in Los Alamos. It was undertaken in narrow, isolated canyons at a considerable distance from the mesa on which the residential quarters and workshops had been built. After Alvarez had completed to his satisfaction, in the spring of 1945, the first development model of this bomb release and tried it out, he handed over production of the final model to Dr. Bainbridge, the Technical Director, and asked Oppenheimer for a new assignment, preferably close to the front line.

At the end of May 1945 Alvarez and his team were sent to the air base on the island of Tinian in the Pacific, from which almost daily raids with ordinary explosive and incendiary bombs were carried out against Japan. He developed there, while awaiting his first definite employment of the atom bomb, a measuring device to be dropped at the same time as the bomb. It was designed for transmission by radio signal to the bombing aircraft of information on the strength of the shock waves released by the new weapon.

Meanwhile Slotin had been busy testing the interior mechanism of the experimental bomb. It consisted of two hemispheres which would come together at the moment of release, thereby enabling the uranium they contained to unite in a "critical mass." The determination of this critical size—referred to simply as "crit" in the Los Alamos jargon—had been one of the chief problems studied by the theoretical department. But the quantity of uranium required, the scattering angle and range of the neutrons to be emitted by the chain reaction, the speed at which the two hemispheres would have to collide and a whole series of other data could be only approximately estimated. If absolute precision and certainty were to be attained, it could only be by way of experiment in every case. Such experiments were assigned to the group in charge of Frisch, the discoverer of fission, who had been brought to Los Alamos from England.

Slotin was one of the members of this group. He was in the habit of experimenting without taking any special protective measures. His only instruments were two screwdrivers, by means of which he allowed the two hemispheres to slide towards each other on a rod, while he watched them with incessant concentration. His object was to do no more than just reach the critical point, the very first step in the chain reaction, which would immediately stop the moment he parted the spheres again. If he passed the point or was not quite quick enough in breaking contact, the mass might become super critical and produce a nuclear explosion. Frisch himself had once nearly lost his life during one of these experiments at Los Alamos.

Slotin knew, of course, by how narrow a margin his chief had escaped death. But the daring young scientist thoroughly enjoyed risking his life in this way. He called it "twisting the dragon's tail." Ever since his earliest youth he had gone in search of fighting, excitement and adventure. He had volunteered for service in the Spanish civil war, more for the sake of the thrill of it than on political grounds. He had often been in extreme danger as an antiaircraft gunner in that war. As soon as the Second World War broke out he immediately joined the Royal Air Force. But in spite of distinguished service in active conflict he was obliged to resign shortly afterwards, when it was discovered that at his medical examination he had managed to conceal the fact that he was nearsighted.

On the way home from Europe to his native city of Winnipeg in Canada Slotin met an acquaintance in Chicago who convinced him that in view of his high scientific qualifications—he had won a prize for biophysics while a student at King's College in London—he could do more for the war effort in a laboratory than in a fighter aircraft. He had therefore found employment first as a biochemist and then as a member of the group which constructed the big cyclotron in the Metallurgical Laboratory of the Manhattan Project. The young man was popular with everyone. Nothing in life seemed to interest him so passionately as his work, to which he devoted himself day and night.

After working at Oak Ridge with Wigner on the development

of new types of reactor, Slotin eventually arrived in Los Alamos. He had hoped to be transferred with Alvarez to Tinian at the beginning of 1945, in order to undertake the assembly there of the explosive heart of the first atom bomb to be used in the war. But as he was a Canadian citizen the security authorities were bound by the regulations to refuse his application. By way of consolation he was given the task of mounting the internal mechanism of the Alamogordo experimental bomb and handing it over officially to the Army on behalf of the Laboratory. A copy of the document sent to him certifying the delivery to the Army of the nuclear component of the first complete atom bomb was thenceforward the principal exhibit in his collection of diplomas, boxing trophies and letters of appreciation of his services.

On May 21, 1946, not quite a year later, Slotin was carrying out an experiment, similar to those he had so often successfully performed in the past. It was connected with the preparation of the second atom-bomb test, to be performed in the waters of the South Sea atoll of Bikini. Suddenly his screwdriver slipped. The hemispheres came too close together and the material became critical. The whole room was instantly filled with a dazzling, bluish glare. Slotin, instead of ducking and thereby possibly saving himself, tore the two hemispheres apart with his hands and thus interrupted the chain reaction. By this action he saved the lives of the seven other persons in the room. He had realized at once that he himself would be bound to succumb to the effects of the excessive radiation dose which he had absorbed. But he did not lose his self-control for a moment. He told his colleagues to go and stand exactly where they had been at the instant of the disaster. He then drew on the blackboard, with his own hand, an accurate sketch of their relative positions, so that the doctors could ascertain the degree of radiation to which each of those present had been exposed.

As he was sitting with Al Graves, the scientist who, except himself, had been most severely infected by the radiation, waiting at the roadside for the car which had been ordered to take them to the hospital, he said quietly to his companion: "You'll come through all right. But I haven't the faintest chance myself." It was only too true. Nine days later the man who had experimentally determined the critical mass for the first atom bomb died in terrible agony.

The recording card of the neutron counter had been left behind in Slotin's laboratory. It showed a thin red line which rose steadily till it stopped abruptly at the instant of the catastrophe. The radiation had become so strong at that moment that the delicate instrument could no longer register it. The scientist charged with ascertaining from the available data what had happened, as a matter of physics, after Slotin's hand had slipped, was Klaus Fuchs.

Strange to say, a terrible fate was also in store for the complement of the cruiser *Indianapolis,* which took to Tinian the greater part of the explosive heart of the first atom bomb destined for use against Japan. Only three men aboard the vessel, the fastest in the American fleet, had any idea what she was carrying. The rest simply supposed that there must be something very important in the big wooden case which had been hoisted aboard, with every precaution, on the morning of July 16 shortly before the ship put to sea. During the voyage from San Francisco to Tinian very special security measures were taken for defense against hostile submarines. Everyone heaved a sigh of relief when the *Indianapolis,* after unloading her secret cargo at Tinian, stood for the offing. But before the cruiser had reached her second port of call she was struck, at five minutes past midnight on July 30, by a torpedo. Owing to a series of unfortunate circumstances news of the sinking did not reach naval headquarters for another four days. Signals from another ship were wrongly taken to be routine reports from the *Indianapolis* of her position and owing to one more misunderstanding she was not reported as overdue at Leyte harbor. Therefore, salvage units arrived too late at the scene of the disaster, and of the ship's 1,196 men only 316 were rescued.

Some days before the first test of the bomb it was an open secret in Alamogordo, even among the wives and children of the

Los Alamos scientists, that some particularly important and exciting event was in preparation. The test was referred to under the code name "Trinity." No clear explanation has hitherto been forthcoming as to why this blasphemous expression was employed, above all in such a connection. One probability is that it was taken from the name of a turquoise mine near Los Alamos, which had been laid under a curse and therefore abandoned by the superstitious Indians. Another guess supposes that it was chosen because at that time the first three atom bombs were approaching completion, and that the code name was derived simply and solely from the existence of that hellish trinity.

The main topic of conversation among the atomic scientists working at Los Alamos naturally turned on the question, "Will the 'gadget' "—the word "bomb" was discreetly avoided—"go off or not?" The majority believed that the theoretical hypotheses would be proved right. But the possibility of failure had always to be taken into account. Alvarez, constructor of the trigger of the bomb, had often enough told over-confident colleagues how in 1943, when his invention for blind landing had been demonstrated to the military authorities, it had failed no less than four times before it eventually worked.

The question whether the first complete atom bomb would be a "dud" or a success, or, as they said at Los Alamos, a "girl" or a "boy," aroused such intense interest that it became the pretext for a charming little game with horror. Lothar W. Nordheim, an atomic physicist who had once been a member of the old guard of Göttingen, relates: "The scientists at Los Alamos had a betting pool before the first test of July 16, 1945, on the size of the burst. But most estimates were too low by far except for one or two wild guesses."

The only estimate that was nearly correct came from Oppenheimer's friend Robert Serber, who had not been in Los Alamos for some time. When he was asked at a later date why he of all people should have been so nearly right in his forecast, he replied: "It was really only out of politeness. I thought that as a guest I ought to name a flatteringly high figure."

On Thursday the 12th and Friday the 13th of July 1945 the components of the interior explosive mechanism of the experimental bomb left Los Alamos by the "back door," along a secret road built during the war. They were transported from "Site S," where they had been assembled, to the experimental area known as the *Jornada del Muerto* (Death Tract) near the village of Oscuro (which means "dark"). Here, in the middle of the desert, a tall frame of iron scaffolding had been put up to hold the bomb. Because of the many thunderstorms experienced during the month it was decided not to place the bomb in position until the last possible moment. A bomb of about equal size, filled with ordinary explosive, had been strung up on the scaffolding a few days before to test the conditions. It had been struck by lightning, and had gone off with a loud bang.

The central portion of the bomb was fitted to it in an old farmhouse under the direction of Dr. Robert Bacher, head of the Bomb Physics Division at Los Alamos. General Farrell, Groves's deputy, writes in this connection: "During final preliminary assembly a bad few minutes developed when the assembly of an important section of the bomb was delayed. The entire unit was machine-tooled to the finest measurement. The insertion was partially completed when it apparently wedged tightly and would go no farther. Dr. Bacher, however, was undismayed and reassured the group that time would solve the problem. In three minutes' time Dr. Bacher's statement was verified and basic assembly was completed without further incident."

Those of the atomic scientists who had not left Los Alamos a week earlier to carry out final preparations held themselves in readiness for departure at any moment. They had with them provisions and also, at the express order of those in charge of the experiment, a snake-bite kit. On July 14 and 15 heavy thunderstorms, accompanied by great quantities of hail, broke over Los Alamos. The participants in the experiment, many of whom then learned for the first time the precise aim and object of the work they had been doing, were addressed by Hans Bethe, head of the Theoretical Division, in the biggest of the community halls, ordinarily used for movies. Bethe's speech ended with the words: "Human calculation indicates that the experiment must succeed. But will nature act in conformity with our calculations?" The audience then boarded buses camouflaged with paint and set off on the four hours' journey to the experimental area.

By two o'clock in the morning all those taking part in the experiment were in their places. They were assembled in the Base Camp, some ten miles from Point Zero where there stood the high scaffolding on which the new, still untested weapon had been placed—the bomb on which they had been working for the last two years and had now finally brought to completion. They tried on the dark glasses with which they had been provided and smeared their faces, by artificial light, with anti-sunburn cream. They could hear dance music from the loudspeakers distributed throughout the area. From time to time the music was interrupted by news of the progress of the preparations. It had been arranged that the "shot" should take place at 4 A.M. But the bad weather rendered a postponement necessary.

At the control point, slightly over five and one half miles from the scaffolding, Oppenheimer and Groves conferred about whether the test should be put off altogether. Groves reports: "During most of the time we were strolling about in the dark, outside the control building, looking up at the stars. We kept assuring each other that either one or both of the two stars visible had grown brighter." After consultation with the meteorological experts it was eventually decided to explode the experimental bomb at 5.30 A.M.

At ten minutes past five Oppenheimer's deputy, the atomic physicist Samuel K. Allison, one of the twenty people in the control room, began to send out time signals. At about the same time Groves, who had by then left the control point and returned to the Base Camp, something over four miles further back, was giving the scientific personnel waiting at the Camp their last instructions. They were to put on their sunglasses and lie down on their faces with their heads turned away. For it was considered practically certain that anyone who tried to observe the flames with the naked eye would be blinded.

During the ensuing period of waiting, which seemed an eternity, hardly a word was spoken. Everyone was giving free play to his thoughts. But so far as those who have been asked can remember, these thoughts were not apocalyptic. Most of the people concerned, it appears, were trying to work out how long it would be before they could shift their uncomfortable position and obtain some kind of view of the spectacle awaited. Fermi, experimental-minded as ever, was holding scraps of paper, with which he meant to gauge the air pressure and thereby estimate the strength of the explosion the moment it occurred. Frisch was intent on memorizing the phenomenon as precisely as possible, without allowing either excitement or preconceived notions to interfere with his faculties of perception. Groves was wondering for the hundredth time whether he had taken every possible step to ensure rapid evacuation in the case of a disaster. Oppenheimer oscillated between fears that the experiment would fail and fears that it would succeed.

Then everything happened faster than it could be understood. No one saw the first flash of the atomic fire itself. It was only possible to see its dazzling white reflection in the sky and on the hills. Those who then ventured to turn their heads perceived a bright ball of flame, growing steadily larger and larger. "Good God, I believe that the long-haired boys have lost control!" a senior officer shouted. Carson Mark, one of the most brilliant members of the Theoretical Division, actually thought—though his intelligence told him the thing was impossible—that the ball of fire would never stop growing till it had enveloped all heaven and earth. At that moment everyone forgot what he had intended to do.

Groves writes: "Some of the men in their excitement, having had three years to get ready for it, at the last minute forgot those welders' helmets and stumbled out of the cars where they were sitting. They were distinctly blinded for two to three seconds. In that time they lost the view of what they had been waiting for over three years to see."

People were transfixed with fright at the power of the explosion. Oppenheimer was clinging to one of the uprights in the control room. A passage from the Bhagavad-Gita, the sacred epic of the Hindus, flashed into his mind.

> If the radiance of a thousand suns
> were to burst into the sky,
> that would be like
> the splendor of the Mighty One——

Yet, when the sinister and gigantic cloud rose up in the far distance over Point Zero, he was reminded of another line from the same source:

> I am become Death, the shatterer of worlds.

Sri Krishna, the Exalted One, lord of the fate of mortals, had uttered the phrase. But Robert Oppenheimer was only a man, into whose hands a mighty, a far too mighty, instrument of power had been given.

It is a striking fact that none of those present reacted to the phenomenon as professionally as he had supposed he would. They all, even those—who constituted the majority—ordinarily without religious faith or even any inclination thereto, recounted their experiences in words derived from the linguistic fields of myth and theology. General Farrell, for example, states: "The whole country was lighted by a searing light with an intensity many times that of the midday sun. . . . Thirty seconds after the explosion came, first, the air blast pressing hard against the people and things, to be followed almost immediately by the strong, sustained, awesome roar which warned of doomsday and made us feel that we puny things were blasphemous to dare tamper with the forces heretofore reserved to the Almighty. Words are inadequate tools for the job of acquainting those not present with the physical, mental and psychological effects. It had to be witnessed to be realized."

Even so cool and matter-of-fact a person as Enrico Fermi received a profound shock, in spite of the retort he had made to all the objections of his colleagues to the bomb during the discussions of the past few weeks. He had always said: "Don't bother me with your conscientious scruples! After all, the thing's superb physics!" Never before had he allowed anyone else to drive his car. But on this occasion he confessed that he did not feel capable of sitting at the wheel and asked a friend to take it for him on the road back to Los Alamos. He told his wife, the morning after his return, that it had seemed to him as if the car were jumping from curve to curve, skipping the straight stretches in between.

It appears that General Groves was the first to regain his composure. He consoled one of the scientists who rushed up to him almost in tears, announcing that the unexpectedly powerful explosion had destroyed all his observation and measuring instruments, with the words: "Well, if the instruments couldn't stand it, the bang must certainly have been a pretty big one. And that, after all, was what we most wanted to know." To General Farrell he remarked: "The war's over. One or two of those things, and Japan will be finished."

The general public was, for the time being, told nothing about this first world-shaking atomic explosion. Dwellers near the experimental area up to a distance of some 125 miles had seen an unusually bright light in the sky about 5.30 A.M. But they were put on the wrong scent by the head of the Manhattan District press agency, Jim Moynahan, who sent the false information that a munitions depot had blown up in the Alamogordo region. He added that no lives had been lost.

On the other hand the security authorities, when they tried to restrict knowledge of the successful test to those who had taken part in it, failed once more in their object. Within a few days the scientists' whispering campaign had carried the news to all the Manhattan Project laboratories. Harrison Brown, one of the younger men on research at Oak Ridge, recalls: "We knew about the fireball, the mushroom cloud, the intense heat. Following Alamogordo many of us signed a petition urging that the atomic bomb should not be used against Japan without prior demonstration and opportunity to surrender. And we urged that the government start immediately to study the possibility of securing international control of the new weapon."

The petition mentioned by Brown had been drafted by Szilard, who, after the failure of his efforts at the White House and the negative results of the Franck Report, had decided to lead a last forlorn hope. His idea was to collect the greatest possible number of signatures from participants in the Manhattan Project protesting against the use of the bomb. When a copy of the petition came into the hands of the director of the Oak Ridge laboratory he at once informed Groves of the movement. It would have been difficult for the General to forbid the men on research to sign the document. He therefore hit on a different method of stopping its further circulation. Szilard's petition was declared "secret." And the law stated that secret papers could only be taken from one place to another under military guard. Thus all Groves had to do was to decree: "Unfortunately, we cannot spare any troops for the protection of this document. Until we can do so it must be kept locked up."

In Chicago the men working in the Metallurgical Laboratory were growing more and more restless. John A. Simpson, a young physicist who took a particularly active part in the efforts to prevent the bomb being dropped, states: "In June an extensive panel discussion was held within the Laboratory by several of the younger scientists, covering subjects ranging from the ways of using the bomb to international controls. Following these discussions the military officials refused to permit more than three people to enter into discussions on the problem at Laboratory meetings. The scientists then resorted to the fantastic technique of holding meetings in a small room, where a succession of about twenty people would, one at a time, enter to discuss these problems with a panel of two or three scientists selected for the evening."

Excitement ran so high in Chicago that eventually the Director, A. H. Compton, had a vote taken, through his deputy, Farrington Daniels, who put the following questions:

Which of the following procedures comes closest to your choice as the way in which any new weapon that we might develop should be used in the Japanese war?

(1) Use them in the manner that is from the military point of view most effective in bringing about prompt Japanese surrender at minimum cost to our own armed forces (23 votes, *i.e.*, 15 per cent).

(2) Give a military demonstration in Japan, to be followed by a renewed opportunity for surrender before full use of the weapon (69 votes, *i.e.*, 46 per cent).

(3) Give an experimental demonstration in this country with representatives of Japan present followed by a new opportunity for surrender before full use of weapon (39 votes, *i.e.*, 26 per cent).

(4) Withhold military use of the weapons but make public experimental demonstration of their effectiveness (16 votes, *i.e.*, 11 per cent).

(5) Maintain as secret as possible all developments of our new weapons and refrain from using them in this war (3 votes, *i.e.*, 2 per cent).

Unfortunately the voting, in which 150 persons participated, took place without any previous debate. Consequently, the greatest number of votes, 69, were cast for the second alternative, suggesting a military demonstration in Japan. But after the first two bombs had been dropped on the center of the town of Hiroshima and on Nagasaki most of the 69 voters explained that they had taken a "military demonstration in Japan" to mean an attack on purely military objectives, not on targets occupied also, in fact mainly, by civilians.

It will be noted from the above summary that, compared with alternative (2), 39 votes were cast for experimental demonstration in the United States, 23 for leaving the military authorities a free hand, 16 for a public demonstration without military use of the bomb and 3 for keeping the whole affair secret and not employing the bomb in the current war.

Szilard had obtained the signatures of 67 prominent scientists before Groves had been able to stop further circulation of the petition. Szilard then sent his appeal direct to President Truman. But he only succeeded, by this procedure, in having the matter again referred to the body by which it had twice previously been held up, *viz.*, the Interim Committee charged with advising the President in this fateful question. The most influential members of that committee in this matter were the four atomic physicists whose task it was to facilitate, as professional experts, the form that advice should take. These were Oppenheimer, Fermi, Compton and Lawrence. For the third time within two months they had the opportunity to throw the weight of their authoritative opinion into the scale. Those who opposed the dropping of the bomb on Japan had good grounds for believing that the four scientists would now, after the Alamogordo test, revise their former judgment. For prior to July 16, no one had known whether and if so with what effect the new weapon would explode. But now all calculations had been exceeded ten and twenty times over. The participants in the experiment no longer spoke of the bomb and its effects as a "firecracker" but as a "shattering experience." Surely so great a shock would induce the committee to plead, at this eleventh hour, for a remission of the death sentence passed upon the prospective victims of the first atomic bombardment ever planned.

The argument which had carried most weight in the informal debates preceding the employment of the bomb was the consideration that although the new weapon would undoubtedly entail the sacrifice of very many human lives, on the other hand it might well prevent even greater destruction of life and property on both sides if it really should bring about an immediate end to the war. Ever since May the American public had remained deeply affected by reports of the exceptionally bloody fighting for the island of Okinawa. Although the Japanese knew that Germany had been defeated and that their own position was now hopeless, they continued to defend themselves with incredible obstinacy and contempt of death. More Americans had been killed or severely wounded on Okinawa alone than during the whole campaign for the reconquest of the Philippines. This fact gave rise to the fear that an invasion of Japan would cost hundreds of thousands of human lives on both sides.

When the four professional experts of the scientific panel set themselves once more to study the crucial problem of employment of the atomic bomb, the following question, Compton recalls, was submitted to them: "Can we think of any other means of ending the war quickly?"

But the dilemma represented by the alternatives of dropping the bomb or allowing the war to go on indefinitely did not, as we know today, correspond with the true nature of the situation. It was based, in exactly the same way as the earlier alternatives—"Either we build an atom bomb or Hitler will do it first"—on a false estimation of the plans and resources of the enemy.

The intelligence services of both the Army and the Navy of the United States were in fact at this date already convinced that the final downfall of Japan could only be a question of a few more weeks. Alfred MacCormack, Military Intelligence Director for the Pacific Theater of War, recollects that "we had such complete control of the air over Japan that we knew when and from what port every ship would put to sea. The Japanese had no longer enough food in stock and their fuel reserves were practically exhausted. We had begun a secret process of mining all their harbors, which was steadily isolating them from the rest of the world. If we had brought this operation to its logical conclusion the destruction of Japan's cities with incendiary and other bombs would have been quite unnecessary. But General Norstad declared, at Washington, that this blockading action was a cowardly proceeding unworthy of the Air Force. It was therefore discontinued."

The surrender of Japan could not only have been achieved by intensification of the blockade. The chances of bringing it about by clever diplomacy were even more favorable. For Japan was at that time more than ripe for capitulation. The country was to a great extent willing to capitulate. At the end of April Fujimura, the Japanese Naval Attaché in the Third Reich, who had gone to Berne when Germany collapsed, was introduced by Dr. Friedrich Hack, an anti-Nazi German, to three close colleagues of Allen Dulles, resident in the Swiss capital as chief of the American intelligence organization, the O.S.S. Fujimura told them he was ready to bring pressure to bear on his government with a view to inducing it to accept American capitulation terms. Almost at the same time the Military Attaché, General Okamoto, acting independently, approached the Dulles organization, through the International Settlements Bank at Basle, with a similar proposal. But both these plans came to nothing, as Washington did not wish to commit itself to a precise statement of terms and Tokyo gave no support whatever to the efforts of the two Japanese in Switzerland.

But another attempt made by Japan for the earliest possible restoration of peace might have been taken more seriously. At the suggestion of the Japanese Emperor a movement was initiated to end the war with the United States through the Soviet Union. The first steps toward that goal were taken July 12—the very day on which the first consignment of components for the Alamogordo test of the atom bomb left Los Alamos. But the Russians had little interest in bringing the war to an end before they themselves entered it against Japan as had been determined at Yalta in the previous February. They accordingly at first did all they could to avoid consultation with the Japanese emissaries. When Ambassador Sato was at last given a hearing, the Russians only passed on Tokyo's proposals to the Americans after intentional delay, making light of their importance.

Washington, however, had long known of these maneuvers, for the Americans had deciphered the Japanese secret code. They had been reading, ever since the middle of July, the urgent instructions sent by radio to Sato in Moscow by Prime Minister Tojo, as well as the replies of the Ambassador. Among other messages they had read the words: "Japan is defeated. We must face that fact and act accordingly."

But Truman, instead of exploiting diplomatically these significant indications of Japanese weakness, issued a proclamation on July 26 at the Potsdam Conference, which was bound to make it difficult for the Japanese to capitulate without "losing face" in the process. The President had at that date already been informed by General Groves, in an impressive report, that the experiment at Alamogordo had succeeded beyond all expectations. The American historian Robert J. C. Butow, who has made a comparative study from both American and Japanese sources of the events that preceded the collapse of Japan, is of the opinion that at this period the war could very well have been brought rapidly to an end by diplomatic measures, for example by conveying the conditions laid down in the Potsdam proclamation to Prince Konoye—who had already been given the widest possible plenipotentiary powers by the Emperor—by the discreet use of political channels instead of broadcasting them to the entire world.

Butow writes: "Had the Allies given the Prince a week of grace in which to obtain his Government's support for acceptance of the proposals, the war might have ended toward the latter part of July or the very beginning of the month of August, without the atomic bomb and without Soviet participation in the conflict."*

But probably the main reason why the American government remained blind to the possibility of such measures was the knowledge that it possessed the atomic bomb. Instead of patiently undoing the knot it appeared more convenient to cut it with a slash or two of the shining new weapon.

It would no doubt have required considerable courage on the part of the responsible politicians and strategists concerned to renounce employment of the bomb for the time being. For they could not help fearing that the entire Manhattan Project, which had hitherto swallowed up nearly two billions of dollars, might perhaps be described, after the war, as a senseless waste of money. In that case the praise and fame they might otherwise expect would turn to mockery and censure.

President Truman, in his memoirs, writes that his "yes" decided the issue of the dropping of the bomb. On this passage General Groves remarked to the author: "Truman did not so much say 'yes' as not say 'no.' It would indeed have taken a lot of nerve to say 'no' at that time."

If even the President of the United States did not dare to

* See Robert J. Butow, *Japan's Decision to Surrender,* Hoover Library Publication No. 24, Stanford University Press, 1954, pp. 133-35.

change gear, any such action was far less to be expected of the four atomic experts of the scientific panel, who had never hitherto offered any serious resistance to the plans of their superiors. They felt themselves caught in a vast machinery and they certainly were inadequately informed as to the true political and strategic situation.* If at that time they had had the moral strength to protest on purely humane grounds against the dropping of the bomb, their attitude would no doubt have deeply impressed the President, the Cabinet and the generals. Once more the four atomic scientists "only did their duty."

The hopes of those who opposed the use of the bomb were revived for a moment longer when they heard that Oppenheimer was closeted with General Groves at a specially arranged meeting. In reality, however, the scientist, in seeking this interview with Groves shortly before the bomb was dropped, merely desired in the first place to convince his companion that it would soon be time to think of constructing less primitive atomic weapons. Thus the sum of a thousand individual acts of an intensely conscientious character led eventually to an act of collective abandonment of conscience, horrifying in its magnitude.†

* General Groves states that he, too, had not been told of the Japanese peace feelers. The State Department, on the other hand, though it knew every detail of the action taken by Tokyo in this connection, had been given not the slightest warning of the imminent use of the new weapon.

† In an interview with Oppenheimer in *Le Monde*, April 29, 1958, he was asked: "At the time you were a member of the Interim Committee and responsible for advising President Truman on scientific questions and the use of the bomb in Japan, did you have the impression that some well-informed persons were able to influence certain decisions made for political reasons?" He answered: "What was expected of this committee of experts was primarily a technical opinion on new questions. Let us not forget that it is to a new government, to men who had not been prepared to exercise the power and still less to resolve the atomic problems, on whom the responsibility of a decision fell in this era. The majority of those who demanded an opinion had not had the time to study the question. On the other hand, President Roosevelt and Sir Winston Churchill showed their complete accord with the fact that the atomic bomb had to be used if it proved necessary to end the war. This opinion weighed heavily in the scale. Unfortunately, time was lacking. It seems possible that a more thorough study of the problem, more prolonged, would have led to those responsible to a more precise or even different conception of what was necessary to do with these new weapons."

Thirteen

The Stricken (1945)

On August 7 at 9 A.M. an officer of the Japanese Army Air Force called at the laboratory of Yoshio Nishina, the best known of the Japanese atomic scientists, which had suffered severe damage in a previous air raid. The officer requested Nishina to accompany him at once to the headquarters of the General Staff.

When Nishina asked what the Staff wanted to see him about, the officer only smiled. Nishina was giving his workers instructions for the work to be done during his absence when a reporter from Domei, the official news agency, arrived. He inquired whether the Professor believed the American broadcast that an atomic bomb had been dropped on Hiroshima.

The scientist was greatly alarmed. Like the overwhelming majority of his countrymen, he had not yet heard of this first atomic bombardment. But he had often enough thought, since 1939, that such a weapon might be constructed and used in war. He had even gone so far as to make private calculations of the amount of destruction it might cause.

The journalist had assumed that the announcement was mere propaganda and hoped that the Professor would confirm that it was a lie. But instead Nishina only nodded, stammering with white lips: "Well, yes—it's quite possibly true. . . ." He then followed the officer who had called to see him.

Nishina was small even for a Japanese. His friendly, almost

quadrangular countenance, spotted with small warts, was known and loved by atomic scientists all over the world. He had worked under Niels Bohr in the twenties and had deduced in Copenhagen, together with another of Bohr's pupils, the "Klein-Nishina formula." On his return to Japan he had founded a school of atomic science. He was the obvious person, therefore, who should be consulted as to the character of the new weapon.

For the first few hours after the Hiroshima catastrophe no one in Tokyo knew what had happened there. The earliest official news was contained in a telegram from the senior civil official of the district of Chugoku. It stated that Hiroshima had been attacked by a "small number of aircraft" which had employed a "wholly new type of bomb." At dawn on August 7 the Deputy Chief of the General Staff, Kawabe, had received a further report with a sentence that was at first incomprehensible: "The whole city of Hiroshima was destroyed instantly by a single bomb."

Kawabe immediately remembered that Nishina had once told him, at an earlier date, that, according to information supplied by the Japanese Naval Intelligence Service, atomic bombardment was a possibility. As soon as the Professor appeared Kawabe asked: "Could you build an atom bomb in six months? In favorable circumstances we might be able to hold out that long."

Nishina replied: "Under present conditions six years would not be long enough. In any case we have no uranium."

The scientist was then asked whether he could suggest any effective method of defense against the new bombs. He was able to make only one suggestion: "Shoot down every hostile aircraft that appears over Japan."

Nishina's statement was too devastating to carry immediate conviction in Tokyo military circles. The same day a hastily summoned "Committee for the study of countermeasures against the new bomb" took the view, based on the opinion of another scientist, that even the masterly technical skill of the Americans was not enough to transport such dangerous apparatus across the whole Pacific, from the United States to Japan.

Nishina had offered to fly to Hiroshima himself and verify his conjectures on the spot. It was decided that a committee formed mainly of military experts should take off that same day, in two planes, from the airfield at Tokorazawa for the scene of the disaster. The plane in which Nishina rode developed engine trouble halfway and returned to Tokyo. There were so few aircraft available in Japan at that time that he had to wait a day before he could start again for Hiroshima.

While the Professor waited, he had an experience which made a deep impression on him. With his pupil Fukuda he was standing in a Tokyo street when a single B-29 appeared in the sky. The inhabitants of Tokyo were used to mass air raids. Since their newspapers had not yet been allowed to give them any information about the new bomb, they paid little attention to this one enemy plane, which had apparently broken away from its formation.

But the two scientists felt like cowards as they ran in search of an air-raid shelter. Fukuda relates:

> At that moment we were both suffering from acute pangs of conscience. We alone knew, as those around us did not, that even a single aircraft with a single bomb might cause a more frightful catastrophe than all the squadrons which had formerly attacked us put together. We wanted to utter a cry of warning to all those indifferent people: "Run for safety! That may be no ordinary aircraft with ordinary bombs!" But the General Staff had strictly enjoined us to keep the secret from the uninitiated, even from our own families. Our lips were therefore sealed. Overcome with rage and shame at not being allowed to warn our fellow creatures, we waited for minute after minute in that air-raid shelter. We scarcely dared to breathe until the "all clear" signal was given. Fortunately no atom bomb was dropped on that occasion. But that temporary piece of luck made no difference to our dejection. As we had not dared to warn our fellow human beings, we felt that we had betrayed them. My revered professor, Nishina, never recovered from the feeling of guilt he experienced that day.

Nishina started next day, for the second time, on his flight to Hiroshima, still hoping that he might after all have been mistaken. In addition to the sorrow he felt as a patriot he was also tormented by the fear that if a scientific super-weapon of this kind had really been constructed and used, the scientists of the West, his friends over so many years, in the eyes of the Japanese people would now be inhuman monsters. When on the afternoon of August 8 his aircraft came within sight of the huge, smoking heap of ruins that had once been a flourishing city, all his fears were confirmed. Later he told the American officers who cross-examined him: "As I surveyed the damage from the air, I decided at a glance that nothing but an atomic bomb could have created such devastation."

The Japanese officers who had reached Hiroshima the day before, headed by Seizo Arisue, Director of the Second Bureau (Intelligence Department) of the Army, still hoped that only an ordinary weapon might have been used. On their arrival the military officer in charge of the airfield had run to meet them. One half of his face was badly burned, the other intact. Pointing to the burns, he reported: "Everything that is exposed gets burned. But anything that is covered even slightly will escape. Therefore it cannot be said that there are no countermeasures."

Other eyewitnesses of the terrible disaster that overwhelmed the city of Hiroshima subsequently described the scenes of human misery. Nishina himself was greatly moved by the magnitude and horror of the spectacle, but he would not allow it to divert him from his work. He remained the outwardly calm professional expert, making precise calculations. His business was to note the measurements, not the sufferings of hell. The fact that the roof tiles of all the houses within a radius of some 650 yards of the point of explosion had melted to a thickness of .004 inches enabled him to compute the enormous temperatures which had developed. The shadows of human beings and objects, retained in the wood of some of the walls—the dazzling light had bleached and scorched everything around them—enabled him to calculate, within a margin of error of only slightly less than 3 per cent, the height at which the bomb had exploded. He even dug up the ground "on the spot right under the point of explosion" in order to test the degree of its radioactivity. Four months later, in December 1945, his whole body developed blotches which he believed to be a delayed result of the examination he had carried out of the radiation still in the debris.

The tireless little man explored the city in all directions, to ascertain the radius over which windows had been shattered by the pressure of the bomb. He visited an anti-aircraft station on the island of Mukai Shima near the city and obtained a description of the attack from the gunners. They told him: "There were really only two B-29's. We can't believe they destroyed the whole city."

On August 10 the various Japanese investigation committees which had spent the last three days trying to reconstruct the course of the disaster met in one of the buildings that were still standing in the neighborhood of Hiroshima. Most of those who attended were now convinced that the Americans had really dropped an atom bomb. One Naval Academy instructor declared that a "different sort of bomb" had been used, probably one containing "liquid air." Nishina unhesitatingly rejected this view. He proceeded to outline briefly the prewar development of atomic research, closing his lecture with the words, "I myself took part in it." The statement sounded like self-accusation, as though he felt that his conduct had been indefensible. The Professor then relapsed into a moody silence of despair, which persisted for a long time afterwards.

One of those who were deeply shocked by the news that an atom bomb had been dropped was the discoverer of uranium fission, Otto Hahn. He could not bear to think that his researches, undertaken without any idea of their practical exploitation, had eventually led to the deaths of tens of thousands of men, women and children. After arrest by the Alsos mission he had been taken, by way of Heidelberg and the American Special Transit Camp, known as the "Dustbin," near Paris, to an English country house at Godmanchester, not far from Cambridge. He was in British custody when he learned of the frightful consequences of the studies he had completed nearly seven years before.

Nine other German physicists were interned at Godmanchester with Hahn. Heisenberg and Weizsäcker were there with some of their group; and Harteck and Bagge, who had worked in Hamburg on Diebner's uranium project; Gerlach, who had been appointed, with the aid of Speer, the Minister of Supply, and in the teeth of opposition by the Party officials, to the post of Supreme Head of Nuclear Physics Research; and Max von Laue, though Goudsmit had assured him that the Allies knew perfectly well he had always been an outspoken opponent of the National Socialist regime.

These ten men had a better time of it, in a material sense, than anyone who had to live in Germany during the months that followed the collapse. They were treated in very friendly, even frankly flattering fashion. The American soldiers themselves who had to mount guard over them during the various stages of their journey noticed that they must be very important people and were fond of guessing who they might be. One of these uniformed guards confided to his prisoner, von Laue: "*You* are Marshal Pétain!"

But the excellent fare and adequate accommodation provided could not relieve the prisoners of their anxiety over the fate of their families, left behind in the chaos of Germany. They had at first been forbidden to correspond directly with anyone in their native land, including their nearest relatives. They had disappeared so entirely from the knowledge of the rest of the world that when the Swedish Academy wished to get in touch with Hahn, who had been proposed for a Nobel prize, he could not at first be found. It was vaguely rumored that he was somewhere in the United States.

Goudsmit, who spent an hour with each of the prisoners before they were transferred, remarks: "Just why these top German physicists were interned in England I never understood. . . . Perhaps our military experts did not know what to do with these scientists after we had found them and felt quite grateful when the British offered to take them over." The scientific head of the Alsos mission explained the secrecy maintained about the place of internment: "All this hush-hush was necessitated by our original assumption that the Germans had the atom bomb or must be close to its secret. As it turned out, they knew practically nothing of significance. But by tracking them down and making such a thorough investigation we might have shown our hand. Actually, the German scientists were sure of their own superiority. It never occurred to them that we might have succeeded where they had failed. But our military security experts could not be sure of this. They could not be sure that if these men were set free the supposition that we might have a gigantic uranium project would become a matter of common knowledge everywhere. The risk was too great. The only thing to do was to segregate the men and keep their colleagues and the rest of the world guessing."

Farm Hall, the secret place of internment for the German atomic physicists, had been built in 1728. Its first owner, a judge named Clark, had been infected by "jail fever" after one of his periodic visits to examine those in custody. He had died of the illness. If in 1945 he had glanced from heaven at his idyllic little seat in the country, where he spent his leisure digging in the garden for Roman coins and fragments of pottery, he would undoubtedly have been amazed to find that his "Tusculan villa" had, for the time being, been turned into a prison. Before the German atomic physicists came to be interned at Farm Hall it had been used as a training school for the British, Dutch, Belgian and French secret agents who were to be landed on the Continent during the German occupation.

It was a large brick house, walled in from the road, with a view of green meadows and high trees. All in all, the Hall was a most agreeable and hospitable place of detention. Two British officers supervised the ten valuable prisoners. Weizsäcker subsequently acknowledged that "these two officers carried out their difficult task of superintending ten discontented physicists with the greatest tact imaginable. We shall always be grateful to them for their behavior." But the discontented ones often regarded their internment as a bit of luck, since it released them for a while from the complex web of duties which scientists, like all others today, are bound to perform. Weizsäcker comments, in recalling the period he spent in the ivory tower which had been created by Army security measures: "If it were not for the constant worry about my family, I should say that I probably never enjoyed myself more than I did then."

Like most of the other prisoners Weizsäcker was able to meditate and write in such peace as he had not known for years. It was at Farm Hall that he worked out some of the most important and striking of his ideas about the origin of the universe. Another of the interned physicists, Max von Laue, wrote a study of Röntgen rays. To keep fit, the sixty-five-year-old Nobel prize winner walked six miles every day. Otto Hahn reports: "This meant that he had to make about fifty rounds of the garden assigned to us, putting a chalk mark on the wall for every round."

Many hours were passed, quite in the style of prisoners of war, getting through the time by playing handball, solving brain-teasers, and rummaging in the library of the house, which was well stocked with old books. Heisenberg read nearly all the works of the British novelist Anthony Trollope. At other times the internees listened to concerts broadcast by the B.B.C. But daily "classes" were also held, at which one of the ten scientists would give an account of his most recent studies. These lectures generally ended with a lively interchange of ideas. All such debates, and also even private conversations and table-talk, were tapped by concealed microphones and recorded on tape. The prisoners learned of this later, because of an accident. One evening, not long before Christmas 1945, they were unexpectedly requested to leave their common living room. It turned out that a soldier who had been installing a loudspeaker for the Christmas party of the troops on guard duty had inadvertently cut the microphone wire in the process.

It would be extremely interesting to listen to the tape recordings today preserved in the secret achives of the British Intelligence Service and hear the discussions which took place among the internees at Farm Hall on the evening of August 6, after the announcement of the dropping of the bomb on Hiroshima. Goudsmit has written a fairly long account of the conversation, but those who took part in it do not consider his report quite accurate. He says that the German experts at first refused to believe the story. "It can't be an atomic bomb," one is said to have remarked. "It's probably propaganda, just as it was in Germany. They may have some new explosive or an extra-large bomb they call 'atomic.' But it's certainly not what we would have called an atomic bomb. It has nothing whatever to do with the uranium problem. . . ." "That being settled"—Goudsmit continues—"the German scientists were able to finish their dinner in peace and even partially digest it. But at nine o'clock came the detailed news broadcast . . . the impact on the ten scientists was shattering . . . they spent hours discussing the science of the bomb and tried to figure out its mechanism. But the radio, for all its details, had not given enough and the German scientists still believed that what we had dropped on Hiroshima was a complete uranium pile. . . ."

Walter Gerlach, who kept a diary, confirms at least one of Goudsmit's statements. He, too, noticed immediately at that time that Heisenberg did not believe in the existence of an American bomb. Weizsäcker makes the following comments on Goudsmit's account:

> Goudsmit himself was not present at the discussions we had, which he describes in so lively a fashion, at our place of internment on the evening when we heard that the atom bomb had been dropped on Hiroshima. He can only have based his narrative on the reports of the two English officers who had been put in charge of us . . . but these officers were not physicists and could therefore certainly not have reproduced accurately the conversations to which they listened about the physics of the bomb. Goudsmit's account therefore contains a number of inaccuracies. In particular, we never supposed that the Americans had dropped a pile. I cannot say, of course,

whether one of us may not have mentioned such a possibility, in the course of the debate, concerned as it was with technical matters which were not clear to us at the time. But if so we should certainly have had no trouble in concluding, from our own technical knowledge, that such an interpretation of the newspaper report would be in the highest degree improbable. . . .

Nor is it true that on receipt of the first broadcast we all soothed ourselves with the reflection that the bomb could not have been atomic. Obviously no one who listens to an excited conversation among ten people can hear everything that is said. It is true enough, however, that we had such a precise knowledge of the difficulties inherent in the production of an atom bomb and considered them so formidable that it had never occurred to us that America would be in a position to produce atom bombs during the war. . . . In our own narrow circle we thought it probable that the United States, if they were to apply their whole resources to that end, would make further progress than ourselves towards a solution of the uranium problem, as well as of other questions. But we considered it improbable that American studies would be promoted in this sense during the war. For the fact was that we underestimated the American potential in supposing that even in the States the actual production of an atom bomb could be practically ruled out. We assumed that in view of this state of affairs the American authorities would decide in advance to postpone such an undertaking until after the war. Nor was our assessment of the position, though quantitatively in error, qualitatively far out. For after all it was not until the war with Germany was over that the bomb was actually completed.

"A very difficult situation," as Gerlach notes in his diary, then arose in the small circle of scientists at the Hall. During the months of internment they had come to be friends. But now the younger men, especially, began to reproach their seniors. Had they been right in not building the bomb? If Germany had possessed such a weapon, would it not have been possible to extort more favorable peace conditions? The older men answered that it was fortunate for German atomic physics to have been spared the heavy burden of guilt which Allied nuclear physicists would probably now have to bear.

Otto Hahn took hardly any part in these heated and often

acrimonious disputes. He was so depressed that his colleagues feared at times he might grow desperate enough to take his own life. "Watch Hahn!" they whispered to one another.

From the diary of Dr. Bagge, August 7, 1945:

> Poor Professor Hahn! He told us that when he first learned of the terrible consequences which atomic fission could have, he had been unable to sleep for several nights and contemplated suicide. At one time there was even an idea of disposing of all uranium in the sea, in order to prevent this catastrophe. . . . At 2 a.m. there was a knock on our door and in came von Laue. "We have to do something, I am very worried about Otto Hahn. This news has upset him dreadfully, and I fear the worst." We stayed up for quite a while and only when we had made sure that Hahn had fallen asleep, did we go to bed.

It was one of the two supervising officers who had first told him about the dropping of the bomb. Hahn had been almost as deeply shocked by the words in which the other had attempted to console him as by the news itself. For when the scientist, who had always condemned Hitler's mania for racial discrimination, exclaimed incredulously: "What, a hundred thousand lives lost? But that's terrible!" his informant replied: "No need to get so excited. It's better for a few thousand Japs to perish than a single one of our boys." *

* The argument that the lives of many soldiers of the Allied Nations were saved by the atomic bombardment of Japan was met at a later date by the well-known American clergyman Monsignor Sheen in the following words: "That was precisely the argument Hitler used in bombing Holland."

Fourteen

The Scientists' Crusade (1945-1946)

The minds of the atomic physicists at Los Alamos had been greatly disturbed and bewildered by the news of the bomb dropped on Hiroshima. O. R. Frisch remembers that one day he suddenly heard loud cries of delight in the corridor outside his study. When he opened the door he saw some of his younger colleagues rushing along with yells of "Whoopee!," like an Indian war cry. They had just heard, over the radio, President Truman reading the report by General Groves of the successful use of the first atom bomb. "It seemed to me that shouts of joy were rather inappropriate," Frisch noted dryly. It was he who, in 1939, had first calculated what enormous energy would be released by splitting the atomic nucleus. That energy had now destroyed tens of thousands of lives.

August 6, 1945, was a black day for people like Einstein, Franck, Szilard and Rabinowitch, who had done their best to prevent use of the bomb. But the men and women up on the mesa were in a quandary. After all, they had worked day and night to achieve their goal. Should they now be proud of what they had done, as it was generally considered they ought to be, in this first moment of surprise? Or should they be ashamed of their work when they thought of the suffering it had caused so many defenseless people? Or again, was it possible—and this position would be the strangest of all, really only comparable with the contradictory data of atomic physics—for one and the same person to feel pride and shame simultaneously?

The whole business became still more confusing when one contrasted the character of this event, so difficult to grasp, with that of the men who had brought it about by the exercise of their intelligence and their deliberate concentration on their effort. In the eyes of the world they had now grown to a stature which no longer corresponded with, in fact contradicted, their true personalities. The godlike magnitude of their performance had given them the standing of mythical figures, more than life size, in the imagination of the public. They were called titans and compared with Prometheus, who had challenged Zeus, the controller of the Fates. They were also called "Devil Gods." But to themselves and their neighbors they seemed the same as they were before, human beings not distinguished for any special virtue or wickedness, contradictory beings in the habit of calculating in business hours, undistracted by "incidental" considerations, their bomb's probable radius of destruction, while in their leisure hours they might be, like Alvin Graves, the most careful of gardeners, rationing their own drinking water to save one of their plants from drying up.

Robert Brode, one of the American physicists who had studied in Göttingen twenty years before, tried to describe his own feelings and those of some of his companions at Los Alamos at that time in the following terms:

> We were naturally shocked by the effect our weapon had produced, and in particular because the bomb had not been aimed, as we had assumed, specifically at the military establishments in Hiroshima, but dropped in the center of the town. But if I am to tell the whole truth I must confess that our relief was really greater than our horror. For at last our families and friends in other cities and countries knew why we had disappeared for years on end. They had now realized that we, too, had been doing our duty. Finally we ourselves also learned that our work had not been in vain. Speaking for myself, I can say that I had no feelings of guilt.

"Willie" Higinbotham, a thirty-four-year-old electronics specialist—the son of a Protestant clergyman and soon afterward prominent among those atomic scientists who felt politically and morally responsible for their work—wrote from Los Alamos, to his mother:

I am not a bit proud of the job we have done . . . the only reason for doing it was to beat the rest of the world to a draw . . . perhaps this is so devastating that man will be forced to be peaceful. The alternative to peace is now unthinkable. But unfortunately there will always be some who don't think. . . . I think I now know the meaning of "mixed emotions." I am afraid that Gandhi is the only real disciple of Christ at present . . . anyway it is over for now and God give us strength in the future. Love, Will.

Some of the atomic physicists at work in Los Alamos knew that the last of the atom bombs—only three had been completed so far—was stored on the island of Tinian, ready for use. In contrast to the bomb dropped on Hiroshima, called the "thin man," this was known as the "fat man." There was every reason to suppose that, with a smaller expenditure, it would be even more destructive. One of the constructors of this last bomb, who for obvious reasons does not wish to be named, admits: "I dreaded the use of this 'better' bomb. I hoped that it would not be used and trembled at the thought of the devastation it would cause. And yet, to be quite frank, I was desperately anxious to find out whether this type of bomb would also do what was expected of it, in short, whether its intricate mechanism would work. These were dreadful thoughts, I know, and still I could not help having them."

Twenty-five atomic scientists and their assistants had meanwhile traveled from Los Alamos to Tinian, under the leadership of Norman Ramsay, to get the fat man ready for use.

So long as no one on the island knew what the "long-haired guys" were really doing in the buildings they occupied, surrounded by a special guard, the military personnel had considered them merely objects for good-natured ridicule. But as soon as the news of the dropping of the first atom bomb became known, they were treated as heroes. There were good grounds for this attitude. For the men of the Marine Corps stationed on the island had learned that they were to bear the brunt, as front-line troops, of the forthcoming landing in Tokyo Bay. But there was now reason to hope that this operation might never take place. A large number of journalists began to arrive at the air base, as well as certain senior officers, who distributed badges to the crew of the *Enola Gay*—the first atom-bomb aircraft, which was named for the mother of the pilot, Paul Tibbetts.

Among the very important persons who visited Tinian at this time was General "Tooey" Spaatz, commander-in-chief of all the Air Forces engaged on that front. Herbert Agnew, one of the atomic experts on the island, relates that "we naturally took him, among other places, to the hangar where we had got the first bomb ready for release. One of my colleagues showed him the little box in which the central mechanism of the bomb had been packed before we fitted it. The General lost his temper. He turned to his adjutant and said: 'You can believe this boy's line of sales talk if you like. But he doesn't pull *my* leg!' The General simply refused to believe that such a little thing had caused such mighty destruction."

It was arranged that certain atomic scientists, including Alvarez, Agnew and the British bomb expert Penney, should accompany this second atomic air raid in another plane. While Alvarez and his friends Philip Morrison and Robert Serber were drinking a can of beer, shortly before starting on the raid, they had a sudden brain wave. They decided to drop a letter with the bomb addressed to their Japanese friend Professor Sagane, with whom they had worked in close contact at the Radiation Laboratory in Berkeley before the war. Three copies of the letter were handwritten in great haste, and one copy securely fastened to each of the three measuring instruments which Alvarez would release over the target. It ran:

Headquarters, Atom-bomb Command.
9 August 1945.

To: Professor R. Sagane.
From: Three of your former scientific colleagues during your stay in the United States.

We are sending you this as a personal message, to urge that you use your influence, as a reputable nuclear physicist, to convince the Japanese General Staff of the terrible consequences which will be suffered by your people if you continue in this war.

You have known for several years that an atomic bomb could be built if a nation were willing to pay the enormous cost of preparing the necessary material. Now that you have seen that we have constructed the production plants, there can be no doubt in your mind that all the output of these factories, working 24 hours a day, will be exploded on your homeland.

Within the space of three weeks we have proof-fired one bomb in the American desert, exploded one in Hiroshima and fired the third this morning.

We implore you to confirm these facts to your leaders and to do your utmost to stop the destruction and waste of life which can only result in the total annihilation of all your cities, if continued. As scientists, we deplore the use to which a beautiful discovery has been put, but we can assure you that unless Japan surrenders at once this rain of atomic bombs will increase manyfold in fury.

One of these messages was found after the bombardment of Nagasaki and handed over to the Japanese Naval Intelligence Division. It was not until much later that it reached the man to whom it was written.

It is not known to what extent this letter contributed to bring about Japan's capitulation. In reality the United States had not a single atomic bomb in reserve, ready for use, at the time the message was dropped. Nor could any fresh bombs be produced for several weeks, possibly for several months, ahead.

The American General Staff had one object in particular in raiding Nagasaki. It was desired to give the enemy the impression that the United States already possessed a whole arsenal of atom bombs and thus induce the Japanese to lay down their arms immediately. The bluff was substantiated, in all innocence, by the message which the three physicists had composed for humanitarian ends. Consequently, even the friendship among scientists of different nations had been misused as a weapon.

Late in the evening of August 11, 1945, the American radio announced: "The United Press has just reported from Berne in Switzerland that the Japanese government has offered unconditional surrender. . . ."

The news caused ecstatic rejoicing at Los Alamos. All contradictory feelings and doubts were for the moment forgotten. Further bloodshed had been prevented by the two "boys" born on the Hill. The war was at an end! A rush was made to extract from their hiding places the supplies of whiskey, gin, vodka and other alcoholic beverages—long since, in expectation of this hour, smuggled into the city of laboratories, hitherto subject to a strict prohibition. People touched glasses happily and drank to peace.

At the climax of one of the many improvised victory parties Professor K., one of the leading specialists in the Explosives Department, rose to his feet, reeling slightly, and dashed out into the night before anyone could stop him. Ever since August 6 he had been working, unknown to all but the security authorities, on a surprise of his own, to be revealed the day the war ended.

A moment later flashes and roars came from all directions. People who rushed out of their houses beheld a magnificent spectacle. The whole of the town of Los Alamos, perched on its precipice, was illuminated by a blinding, shimmering glare. The towering red rocks glowed in the reflection of the flames. Arrowy fountains of sparks shot up out of the canyons. There semed no end to the bangs, loud reports and thunderous echoes. Professor K. had connected by wire two or three dozen small munitions dumps at concealed spots; by pressing a button they would explode.

After the victory fireworks had burned themselves out and only occasional belated explosions could be heard as an aftermath to the main display, people returned to their houses and began to listen in again in the hope of hearing more details of the surrender. They learned, however, that the news of Japan's capitulation had unfortunately been premature.

Four days later came the announcement that Japan really had surrendered. This time there was no rejoicing at first, but after a while, despite the late hour of the announcement, a victory parade was organized at Los Alamos. It was led by a jeep with more than a dozen of the younger scientists clinging to it. The slim figure of Willie Higinbotham was seated on the shoulders of the driver. He played lively tunes on his accordion and banged a kettledrum made of the lids of two dustbins, to make sure that those who happened to be asleep should have no doubt that peace had broken out.

Lights went on again in most of the houses. Dorm parties began in the bachelor sleeping quarters. Dancing went on until dawn. The staff were excused from work on the following day. So it continued for two days and two nights.

But when the rejoicings came to an end it was found that for the present everything was to go on as before. The world might be under the impression that peace had come again. But so far as the people on research at Los Alamos, Oak Ridge, Hanford and Chicago were concerned, the same strict rules of secrecy prevailed as had been in force during the war.

The younger workers on the Manhattan Project, in particular, found these conditions unbearable. They began to grumble. A typical complaint came from Herbert Anderson, a young American physicist. He had taken part in Fermi's first uranium experiments at Columbia University, during which he had contracted lifelong beryllium poisoning. Shortly after the war Anderson wrote to a friend: "We ought to resist every encroachment upon our rights as human beings and citizens. The war has been won. We wish to be free again."

These scientists were not only concerned about their personal freedom. They desired in particular to be free to enlighten their fellow men about the terrors of the new weapon. When they read in the newspapers, at that time, that members of Congress were in favor of the United States keeping the secret of the atom bomb to themselves, the physicists would have liked to retort that there was no atomic secret which could not be detected within a very short time by any nation scientifically of the first rank. They would have liked to press for the immediate convocation, on American initiative, of an international conference on the control of atomic development, as had been desired by Bohr, Szilard and the author of the Franck Report.

A special subject brought up by the scientists at Los Alamos was the game of hide-and-seek played by the Army with the problem of radioactivity. Even before the atomic weapon had first been used some physicists had entreated General Groves to allow pamphlets to be dropped at the same time as the bomb, pointing out the unfamiliar dangers of radioactivity arising from the explosion of this new weapon. This request had been refused by the military authorities, for they feared that such warnings might be interpreted as a confession that they had been employing a type of weapon like poison gas.

They proceeded, probably from similar motives, to try to divert attention from the radioactive effects of atomic bombardment. It was explained that there was now no dangerous radioactivity to be found in the ruins of Hiroshima, and the number the victims who had been exposed, at the moment of the explosion, to a fatal dose of radiation or one likely to cause chronic illness, was kept secret. Groves stated openly at a Congressional hearing that he had heard death from radiation was "very pleasant."

Such observations made the Los Alamos scientists' blood boil. For at that very moment their twenty-six-year-old colleague Harry Dagnian was struggling against the menace of a cruel death from the effects of radiation.

On August 21, 1945, during an experiment with a small quantity of fissile material, Dagnian had set off a chain reaction for the fraction of a second. His right hand had received a huge dose of radiation. After admission to hospital within half an hour of the accident, the patient had at first noticed only a certain loss of sensation in the fingers, occasionally superseded by slight tingling. But soon his hands grew more and more swollen and his general condition deteriorated rapidly.

Delirium set in. The young physicist complained of severe internal pains, for it was now that the effect of the gamma rays, which had penetrated far beneath the skin to the interior of the body, began to be perceptible. The patient's hair dropped out.

The white corpuscles of his blood increased rapidly. Twenty-four days later he died.*

For the first time death by radiation, which the men of Los Alamos had inflicted upon thousands of Japanese by constructing their weapon, had overtaken one of themselves. For the first time the dangerous effects of the new power had been brought close, not in the form of a distant statistic, but as the suffering, pain and fatal sickness of one of their own group.

The accident to Henry Dagnian intensified the movement which had begun in all the atomic laboratories among those scientists who intended to tell the world the whole truth about the new weapon and entreat their fellow men to renounce all use of atomic energy in warfare. Nine days after Dagnian had been taken to the hospital shed on the Hill, the Association of Atomic Scientists, headed by Higinbotham, was formed in Los Alamos. About a hundred of the men in research immediately joined it. Similar groups had already arisen in Chicago, at Oak Ridge and in New York. The groups got in touch with one another and came to a common decision to enlighten the public and thus bring strong pressure to bear on the statesmen of the country, in spite of the fact that such an appeal would constitute an infringement of the Army regulations to which the members of the Association were still subject. Such was the start of the movement which later became known, in a somewhat exaggerated phrase, as the "revolt of the atomic scientists."

Seldom can jubilation have made a man so sad and adulation made a man so skeptical as they did Robert Oppenheimer as he watched the frenzied delight with which his countrymen greeted the end of the Second World War. He, known only to a small

* Exactly eight months after this first accident came the one which befell Louis Slotin—described in Chapter XII. As it was considered absolutely essential to keep this affair a secret, residents of Los Alamos were even forbidden to decline invitations to a reception arranged a long time in advance in honor of Santa Fe notabilities, who had been asked to visit the Hill. Even some of Slotin's closest friends, for example Philip Morrison, were obliged to appear at this cocktail party, in between attendances at the bedside of the dying man, and behave as though they had not a care in the world.

circle of his scientific colleagues and a handful of politicians, had suddenly come to be an object of mass admiration. As the alleged "father of the atom bomb"—it was a designation he always repudiated as oversimplified—the learned physicist was saluted on all sides as though he were a victorious commander-in-chief. He was regarded not only as the man whose miraculous weapon had spared the country the dreaded prospect of heavy casualties in an invasion of Japan and another winter of war, but also as a new kind of peacemaker, whose amazing discovery would make all armies and wars superfluous from this time onward.

Oppenheimer, however, knew too much to be able to acquiesce in this overwhelming tide of optimism about the future. He must at that time have observed all those who were not in the picture and showed such enthusiasm for the coming paradise of peace with the same sadness with which adults sometimes watch the innocent play of children.

When Oppenheimer speculated about the future, his mind was overshadowed by two complex sets of facts. In the first place, it was clear to him that the two atom bombs which had been dropped on Hiroshima and Nagasaki did not represent the height or even an extreme limit, but only the beginning, of a new kind of weapons development whose limits could still not be seen. Even before the completion of the uranium bomb he had written two letters dated September 20, 1944, and October 4, 1944, to a friend, Professor Tolman, chairman of a research committee constituted almost a year before the end of the war to study the future of atomic energy, pointing out that because of wartime conditions they had been able to produce only a relatively primitive atomic weapon. These had been his words:

> Whatever technical superiority this country may at present possess in dealing with the scientific and technical aspects of the problem of the exploitation of nuclear reactions to produce explosive weapons has resulted from a few years of work which was, to be sure, intensive, but inevitably badly planned. Such superiority can probably only be maintained through continued further development of both the technical and the underlying scientific aspects of the prob-

lem. For this purpose both the availability of radioactive materials and the participation of qualified engineers and scientists are equally indispensable. No government can adequately meet its responsibilities for defense if it rests content with the wartime results of this project.

In the second place Oppenheimer knew from personal experience—the degrading interviews forced upon him in 1943—that the germ of atomic rivalry between the two great powers, the United States and the Soviet Union, then still allies, already existed. Unlike his military chief, General Groves, who believed that it would be ten, twenty or even sixty years before the USSR could develop its own atom bomb, Oppenheimer had a high opinion of Soviet research. His views had been quite recently substantiated by Irving Langmuir, an American who had won a Nobel prize for chemistry, on his return from Moscow, where he had been the guest of the Academy of Sciences. Langmuir had no doubt that the Russians could, if they wished, construct atom bombs within a relatively short time and might well have done so already. He even considered that the Soviet Union, as a totalitarian state, could easily initiate a bigger program of atomic armaments than would be possible for the United States.

Such considerations of practical politics at first prevented Oppenheimer, the atomic physicist whose public prestige probably stood higher just after the war than any other, from raising his voice to join in the steadily increasing chorus of warnings. While men like Einstein, Szilard, Franck and Urey talked of the need for an understanding with Russia, Oppenheimer was at the very same time trying to arrange for patrols of aircraft furnished with sensitive measuring instruments to detect any atomic test explosion that might take place in Russia or anywhere else in the world. During the actual week in which the first two atom bombs were dropped, Oppenheimer, Compton, Fermi and Lawrence had already laid down the lines on which future atomic armament should proceed. Oppenheimer himself strenuously opposed the growing tendency of scientists and also of many government officials to "hand back Los Alamos to the desert foxes." In personal conversations and public speeches he endeavored, usually with success, to persuade his collaborators to remain at Los Alamos for, at any rate, some time longer. He felt himself more than ever responsible for this extraordinary settlement "on the edge of the world." His persuasive ability and diplomatic skill gained him new friends among the soldiers stationed at Los Alamos. They had expected a special public citation by the President as a reward for their services. When it failed to materialize they grumbled and protested. Oppenheimer learned of their discontent; he wrote a personal letter of thanks, signed with his own hand, and had a copy delivered to each man. This step made him more popular than ever with the G.I.'s.

On the other hand Oppenheimer began to lose more and more friends among his closest colleagues, who, with few exceptions, had idolized him for years. They had hoped that he would now act as their spokesman to the world, since they themselves were still sworn to secrecy. But whenever they approached him, he invariably replied: "Patience, patience! Just now delicate questions as to the future control of atomic energy are being discussed. We scientists must be careful not to rock the boat. We mustn't interfere."

The delaying answers Oppenheimer gave to the worried young scientists of Los Alamos and also to those of Oak Ridge, when he paid that establishment a visit, resembled the advice proffered by A. H. Compton, head of the Metallurgical Laboratory in Chicago, to the scientists of the Laboratory. He repeated again and again: "Don't take any action. If you do, you will endanger important political developments." It seemed clear that he could only be referring to secret negotiations with Moscow. So the scientists held their tongues, as Compton recommended.

But towards the end of September the news filtered through that no conversations whatever with the Russians on atomic problems had yet been initiated from the American side. At a cabinet meeting on September 21 the American government, with the exception of the former Vice-President and present

Secretary of Commerce, Wallace, had decided for the time being against any revelation of atomic secrets, regarded as a sacred trust. What, then, could Compton have meant? Szilard determined to find out. It was due to his pertinacity that the scientists eventually discovered the truth at which Oppenheimer and Compton had only hinted. Conversations relating to the control of atomic energy had in fact taken place in Washington. Only they had not dealt with international control, as had been supposed, but with the form of control to which the new power was in future to be subject in the United States.

Almost every scientist at that time was of the opinion that there ought to be some sort of public supervision of atomic energy. Now, for the first time in history, something had been invented which in irresponsible hands might imperil the lives of all citizens of the state and perhaps of the entire population of the globe. But everything depended upon *who,* in the name of the nation, would exercise such control. Should direction of the new atomic industry be placed, as in time of war, in the hands of the military authorities?

Szilard gathered from Compton that some such plan was in view. The latter also revealed to him, under pressure, that the War Department, which had framed the new legislative proposals for the control of atomic energy, considered it most important that the bill should pass both houses of Congress without difficulties and also, if possible, without debate.

At this news Szilard lost his patience. He went straight to Washington in order to find out what exactly this bill, hitherto so anxiously kept secret from all the world, might contain. Bob Lamb, C.I.O. representative in Washington, procured him a copy of the bill. Szilard was greatly agitated at what he read. His negative reaction to the bill's contents was supported by the legal faculty of his university in Chicago when he submitted the document to that body. If such a law were to be passed by the representatives of the American people, all future developments in atomic research, instead of being at last directed to the peaceful exploitation of this rich source of energy, would be for the most part misused for the purposes of armament. And yet atomic scientists were supposed to comply with the extremely strict secrecy regulations applicable to them, under the threat of long prison sentences for infringement! If the bill became law the result would soon be, as Chester Barnard, a director of the Rockefeller Foundation, had exclaimed with misgivings when he first heard of the atom bomb, "the end of democracy."

The scheme had been ingeniously contrived. The War Department had drafted the new legislation under the direction of Kenneth Royall, the Assistant Secretary of War, and with the help of General Groves. The Department managed to introduce it as unobtrusively as possible into a Congress overburdened with urgent bills. But under the Constitution public hearings were required before any new legislative proposal was read and debated in Congress. At such hearings qualified supporters and opponents of the bill expressed their opinions. Congressman Andrew May, a small-time attorney from Kentucky, whose many years in the House of Representatives had secured him the chairmanship of the Military Affairs Committee, succeeded in holding hearings on the bill—which he and Senator Johnson of Colorado were to introduce—without any publicity. Only four people had been asked to testify on behalf of the bill. They were the Secretary of War, Patterson, and General Groves, who were both naturally in favor of it, together with the scientists Vannevar Bush and James Conant, who had both collaborated as consultants in the formulation of the bill.

It was only after Szilard, at the last minute, had alarmed his colleagues, that May was compelled, by the pressure of public opinion aroused by statements from the scientists, to arrange further hearings, to be attended by well-known opponents of his bill. One can imagine his irritation with Szilard, who instantly presented himself as the first witness against the proposed legislation.

It was just six years since Szilard, on his way to Einstein's summer home, had doubted whether he ought to continue with his fateful mission. What he had foreseen then had come true. The military authorities had no intention of relaxing their control of the new source of energy. And he himself, for daring to oppose them, was now treated, in spite of his contribution to the development of the new power, worse than a defendant. Congressman May, who presided over the hearing, tried in every possible way to provoke and confuse the scientist. He pretended he had not caught Szilard's name properly, or could not pronounce it, and persistently called him "Mr. Sighland." Szilard talked for an hour and forty minutes, and was continually interrupted and deliberately misunderstood. He was rudely called to order for not answering intricate questions with a plain "yes" or "no." The witness was also told repeatedly he was taking up too much of the hearing's valuable time.

Szilard, by nature a temperamental man, controlled his indignation with remarkable restraint. He saw through the traps laid for him. He allowed neither taunts nor accusations to disconcert him and eventually convinced most of the members of the committee he addressed that his resistance to the continuance of military control of the development of atomic energy was well grounded. He thus won the first skirmish in the month-long struggle of the atomic scientists to ensure civil control. His adversary, Congressman May, apparently so devoted to the interests of the military authorities, was soon afterwards forced to retire from public life and serve a prison sentence for showing favors to an industrialist who had got Army contracts by corrupt practices.

As soon as copies of the May-Johnson Bill reached the atomic laboratories and the universities, the members of the new scientific associations, mostly the younger generation of scientists, determined to send delegates to New York and Washington. They were anxious to enter the political arena to campaign for more satisfactory legislation for the control of atomic power. By the middle of November the local groups had combined into a single body, the Federation of Atomic Scientists. The word "atomic" was later replaced by "American," for a great many of the members had nothing whatever to do with nuclear research. But at that time, in the autumn of 1945, the ominous adjective was still indispensable. It was still a name to conjure with. All doors opened at the word "atomic," the new superlative. Senator Tydings, for instance, declared that an atomic scientist is one of the "few persons whose intellectual development in many respects—and especially in the scientific field—bears the same relation to that of the rest of us as a range of mountains bears to a molehill."

The atomic scientists had become important people. That was their first discovery when they returned from their laboratories to the world at large. "Before the war we were supposed to be completely ignorant of the world and inexperienced in its ways. But now we are regarded as the ultimate authorities on all possible subjects, from nylon stockings to the best form of international organization," one of them remarked with mildly ironic self-detachment, after he had become somewhat accustomed to being dazzled by flash bulbs and confronted by microphones and newsreel cameras.

The more sensitive of these scientists suffered increased pangs of conscience when they realized, as the biologist Dr. Theodor Hauschka put it in a bitter open letter to Oppenheimer, that their prestige came chiefly because they had been "brilliant collaborators with death." But whenever they started to confess their "sins" public interest in them increased. Those who unburden their hearts can nearly always count on a sympathetic audience which not only pardons but admires them. Many of the scientists very soon perceived that this asset of accumulated attention and respect might, perhaps, be converted into the current coin of a genuine political influence. They accordingly began "the last Crusade," as their efforts were called by Michael Amrine, an idealistic young writer who placed himself at their disposal in those days. It was a crusade undertaken by men who were children in political affairs and yet—or possibly for that very reason—gradually made headway in Washington against cunning politicians and apparently invincible vested interests.

Amrine, the loyal historian of this unusual movement, describes the mood that inspired it:

These men had rediscovered their personal, human consciences and were determined to overcome all opposition in order to guide society back to the road of progress and divert it from that which led to annihilation. The manifesto in which they announced this aim was a small sheet of paper written in single space on each side. A radio reporter remarked later that it seemed to have been duplicated with a wet handkerchief. He could not have known, of course, that the scientists only possessed an office which had been lent to them on the fourth floor of a house without an elevator. They had only one room, where there were not enough tables and chairs, so that world-renowned Nobel prizewinners and students had to squat on the floor while they passed to one another the statements and petitions which were subsequently heard by the entire world.

Such was the beginning of an amazing campaign carried out in the face of indifference from the White House, the State Department and Congress, and against powerful and well-organized opposition. Experienced people in Washington all shook their heads. They warned the "league of frightened men," as the scientists were called, not to expect their undertaking to succeed.

During the winter of 1945 the scientists' vision of a new world without hunger or cold was being jotted down by men in thick overcoats in an unheated office immediately above Larry's Coffee Shop on L Street. These men learned the language of politics with surprising rapidity. For instance, they first wrote: "The transformation of mass into energy, as understood by us, has fundamentally changed the nature of the world, as hitherto conceived by us." But that was much too abstract and cautiously formulated a sentence to make any impression. Soon afterwards they were addressing politicians in the following jazzed-up terms: "Senator, if a single one of the new bombs were to burst on the railway station at Washington, the marble on top of the Capitol here would be ground to powder. You yourself and most of your colleagues would probably be dead within the first few minutes." That was effective.

What the young scientists lacked in political experience they made up for by an enthusiasm and sincerity which deeply impressed the politicians and in particular the representatives of the press in Washington. It was known that this strangest of all lobbies was financed only by voluntary contributions from the scientists and that many of them who had been given no leave for years were now dedicating their first free time to this public question. That they were really indefatigable is proved by the entries in a gray-covered, oblong logbook in which every scientist working for the Federation wrote down his doings at the end of the day.

Atomic scientists were the first to enter the anterooms of congressmen early in the morning. Later on they visited editorial offices to distribute the statements which they had themselves typed and duplicated. At noon they gave lunchtime lectures to all sorts of societies—answering such questions as "What color is plutonium?" In the afternoons they sometimes even ventured into the lions' den itself, the Army Hospital, or attended the teaparties given for them by Mrs. Pinchot, a politically influential member of Washington society. Late in the afternoon they were to be found at cocktail parties where they might meet important persons. Some also conducted evening classes in nuclear physics for congressmen and government officials. Others discussed their missionary task, far into the night, with doctors, sociologists and representatives of the Church, the press and the film world.

The first result of all these activities was a substitute for the May-Johnson Bill, framed by the scientists in collaboration with Senator McMahon and now laid before Congress. The next problem was an additional rider attached to the bill by Senator Vandenberg, who had taken this indirect method of smuggling in military control again. The scientists contrived to have this smothered under an avalanche of letters of protest from 75,000 indignant voters. At last, in July 1946, when the McMahon Bill, which handed over control of atomic research development in the United States to a civil commission, was made law, the scientists were able to taste the fruits of victory.

But that victory very soon turned out to have been a Pyrrhic one.

Fifteen

The Bitter Years (1947-1955)

In October 1945 Robert Oppenheimer announced his resignation as Director at Los Alamos. His decision caused much astonishment among the many atomic scientists who had stayed on the Hill. For Oppie had opposed, both in his public speeches and in private conversation, the view prevalent at the end of the war among most of his colleagues that they should now return as soon as possible to the investigation of basic principles for peaceful purposes, because armament research had not after all, despite its practical results, led to the discovery of any important new data in the field of nuclear physics. Teller, who had always admired Oppenheimer but had never been able to get on with him personally, now seized the opportunity to call his attention to this inconsistent attitude. "Three months ago," he complained, "you told me that by all means I should stay. Now you tell me I should leave."

Oppenheimer had said that he proposed to devote himself for the future chiefly to the resumption of his former teaching activities in Berkeley and Pasadena. Such may actually have been his intention at one time. But the war years had changed him. He had become an extremely successful organizer, a planner and a politician of high caliber. Back in 1935 he had explosively and arrogantly refused the request of one of the best-known science reporters, William L. Laurence of the New York *Times,* for an explanation of some of his scientific feats that could be understood by the layman. But now he had a masterly understanding of how to handle his public relations. Oppenheimer typified for the general public the new worldly variety of scientist, with the mighty forces of Nature arrayed behind him, just as the generals commanded their divisions and the politicians their masses of constituents.

He was more and more to be found in government offices, less and less in the lecture room. He had become the oracle of diplomats and strategists. A new stage had begun in the career of this extraordinary man. The fact could even be noted in his altered appearance and behavior. He now wore his graying hair cut very short—as if to prove even by this detail that he was no longer one of the "long-haired." His movements had a military abruptness. His voice could cover a whole range of tones, from deliberately arrogant through judiciously reflective to irresistibly demonstrative of warmth of feeling. He was regarded as a "scientific statesman," with an important influence on the decisions taken in great public issues, as the "Gray Eminence" of the State Department and the Pentagon, and simultaneously as the preceptor of those in power, capable of turning their offices into classrooms from one moment to the next, when he would stand at a blackboard trying to teach them the elements of nuclear physics. He probably considered himself not only their tutor but also their spiritual director.

Oppenheimer's friends, however, believed that Washington's influence on him was greater than his influence on Washington. They were annoyed to find that, despite his private criticism of the May-Johnson Bill when he was among his scientific colleagues, his public statements favored the bill on "tactical grounds." "Better a bad law than no law at all" was his argument.* It was also held against him that, although he had remarked that the physicists were now aware that they had "known sin," he made no suggestions as to how they might show their remorse in a practical form. Again, while he did undoubtedly take a leading part in drafting the plan supported by the scientists for the control of atomic research, which the

* Oppenheimer had proposed, at the hearings on the bill, that General Marshall should be made chairman of the Atomic Energy Commission.

Americans intended to submit to the United Nations, at the same time he told the generals and politicians privately that he considered the proposals went much too far to be really acceptable to the Russians. There was no need, he hinted, for any great uneasiness about the matter.

The physicists had the impression at this time, when they met Oppie, that he was no longer quite one of them. Some were no doubt susceptible to the glamour that now surrounded him, but it was his best friends, in particular, who grew cold towards him. One of Oppenheimer's former favorite pupils relates: "When Oppie started talking about Dean Acheson as simply 'Dean,' and actually referred to General Marshall, as merely 'George,' I knew that we did not move in the same circles any more and that we had come to the parting of the ways. I think that his sudden fame and the new position he now occupied had gone to his head so much that he began to consider himself God Almighty, able to put the whole world to rights."

After Oppenheimer had left Los Alamos a general exodus began. Huge removal vans were to be seen descending the wide new road down into the valley. They carried furniture, trunks and all sorts of local souvenirs, from Indian bracelets to a full-grown saddle horse which Herbert Anderson was taking back to Chicago with him. General Groves remained boss of the research laboratories pending the passing by Congress of a new law governing the control of the installations. He complained: "My first and second teams have left. My third and fourth teams are leaving."

At this period, in February 1946, David Lilienthal, later chairman of the Atomic Energy Commission, paid a visit to Los Alamos in his capacity as adviser to the State Department on the preparation of the American plan for the international control of atomic energy. He found the settlement in a rather neglected condition. He reported:

Deterioration had set in, as one might expect. Scientists had left the project in large numbers. Contractors had declined to go forward, such as Du Pont. Du Pont turned in its contract at Hanford. There was great uncertainty. Morale was badly shot. At Los Alamos we found the most serious situation because although some very able men remained, the top management of that project had left for the universities. We found a great many health hazards and fire hazards that were very damaging to morale . . . there was no inventory of the properties. There was no accounting. This whole thing had been done so hastily that it had not been possible to do that. These things made it very difficult for the men who were operating to make head or tail of what they were doing. The net effect of that was a very depressed state of mind.

In order to raise the spirits of residents on the Hill a little, Oppenheimer's successor, a former professor of physics and officer in the Naval Reserve named Norris Bradbury, sent for a famous jazz orchestra and a wrestling team. The wrestlers, in particular, were a great success. The Los Alamos *Times,* which began publication after the war, reported that they were vociferously encouraged with shouts of "Bite him, bite him! Tear his hair! Don't worry about your children! We will take care of them!" But this outlet apparently did not suffice to relieve the Hill residents of their bottled-up aggressiveness. So the psychiatrist Dr. John Warkentin was summoned to Los Alamos to treat their neuroses.

In addition to the usual shortage of water, the neglect of roads, hedges and open spaces and the dilapidation of the houses, which had been hastily constructed with green wood, the shift in public opinion became noticeable, even on these remote heights. The changed mood had been brought about under the influence of the "scientists' crusade," the horrified reports of eyewitnesses from Hiroshima and essays such as the famous article by Norman Cousins—which he afterwards developed into a short book, *Modern Man Is Obsolete.* It was now considered old-fashioned or even disreputable still to be working on atom bombs.

The Bomb That Fell on America, an epic in verse by the poet Hermann Hagedorn, represented so vividly the deepest feelings of many Americans that it went into a dozen editions in a few months. One passage read:

> When the bomb fell on America it fell on people.
> It didn't dissolve them as it dissolved people in Hiroshima.
> It did not dissolve their bodies.
> But it dissolved something vitally important to the greatest of them and the least.
> What it dissolved were their links with the past and with the future.
> There was something new in the world that set them off forever from what had been.
> Something terrifying and big, beyond any conceivable earthly dimensions. . . .
> It made the earth that seemed so solid, Main Street, that seemed so well paved, a kind of vast jelly, quivering and dividing underfoot. . . .
> What have we done, my country, what have we done?

The profound shock to people's feelings of security is illustrated by the retort, published in *Time,* of an eight-year-old boy to the question: "What do you want to be when you grow up?" He answered: *"Alive!"*

But the "realists," as all those people called themselves who favored retention of the secret of the atom bomb by the United States alone and were already preparing a commanding position for the new weapon in the American Armed Forces' arsenal, worked on at their armament plans undisturbed by the change in public opinion. In September 1945, less than a month after the end of the war, ground began to be cleared for a new atomic-bomb factory near Albuquerque, at the foot of the Sandia Mountains, not far from Los Alamos. The bombs to be manufactured here would be mass-produced without the attention to detail which had been devoted to the original prototypes.

The exodus of the atomic scientists from the armaments laboratories did not worry General Groves quite so much as it had orginally, once he had got over his first disappointment. He was sure his "little sheep" would find their way back.

Meanwhile German armament technicians, despite the protests of American scientists, were being imported to the United States. Those recruited in this way—the affair was known as "Operation Paperclip" to the military authorities—were mainly members of the research departments of the German Air Ministry and constructors of "reprisal weapons." No notice was taken of their former political views. At a time when Americans in occupied Germany were not even allowed to shake hands with people who had been Hitler's enemies or had obeyed him against their wills, a number of those who had openly professed Nazism and had worked on the V2 rockets and other instruments of destruction were invited to the United States to assist in the American armaments industry.

But protests like Hans Bethe's against such a peculiar choice of personnel were disregarded by the Armed Forces. Their attitude was that if they themselves did not get hold of these scientific brains the Russians soon would. As a matter of fact the Red Army, with an ideological nonchalance that rivaled that of their Western allies, really had taken into custody, as war booty, a large number of scientific and technical specialists.

Even the Americans, who later on gave them regular contracts, used somewhat rough methods in their hunt for German scientists immediately after the end of hostilities. For example, the American military police, months after the Armistice, seized a certain "atomic scientist" in Bremen. He was transported, in spite of his desperate protests, to the United States. There he was cross-examined, day after day, about his knowledge of nuclear physics. But unlike most of the other German scientists, who had reported willingly enough to their new masters on their war work, this captive proved to be frankly obstinate. He stubbornly maintained that he knew nothing of atomic research except what he had read in the newspapers, and was, in point of fact, a tailor. He was believed to be shamming until someone hit on the notion of handing him a needle and thread. He dumbfounded his warders by doing really excellent work on their shirts and trousers. It turned out in the end that the man had only been carried off across the ocean because his name was Heinrich Jordan. The M.P.'s had mistaken him for Pascual Jordan, the famous theoretical physicist and former pupil of Max Born.

Another mistake made by the Army authorities could not so easily be put right. In conformity with an order which was supposed to have been issued in General Groves's office, a detachment of the American Army of occupation in Japan, under the command of a Major O'Hearn, destroyed Professor Nishina's two cyclotrons, in the erroneous belief that they might possibly serve for the production of atomic bombs. Before the forceful complaints of the Japanese scientist reached his colleagues in the United States, the demolition unit of the Eighth Army had already, in five days and nights, completed the work of destruction. All the petitions forwarded by the American physicists, who compared this act of vandalism with Hitler's burning of books, came too late.

The American scientists' strongest protests were leveled at the plan for carrying out atomic-bomb tests in the summer of 1946. They took the view that such an experiment would be regarded by public opinion in the rest of the world as "saber-rattling" and would be bound to prejudice negotiations for international control. The atomic maneuvers had been proposed by the Navy, which declared that they were a necessary preliminary to the new naval building program, future naval strategy and naval security measures against the new weapon. The Federation of Atomic Scientists and many other scientific speakers objected that neither scientific nor strategic importance could be attributed to the new tests, which were to be carried out off the atoll of Bikini. In case of war, they pointed out, an enemy would use such expensive weapons, not against targets as difficult to hit as battleships but against large cities, where they would be bound to cause a maximum of devastation. It was forecast that the public would obtain an entirely false idea of the power of the new weapon from the tests proposed.

The experiments at Bikini were postponed for a while; they were considered a discordant accompaniment to the forthcoming presentation of the American plan for international control to the United Nations. But in July 1946 they were held. Their material effect was, as the professionals had prophesied, surprisingly small. But their spiritual effect was great. They soothed the fears of the American public almost as much as the bombs dropped on Japan had aroused them. William Laurence, the only American journalist permitted to attend both the experiment at Alamogordo and the atomic bombardment of Nagasaki, wrote at that time:

> On returning from Bikini one is amazed to find the profound change in the public attitude toward the problem of the atomic bomb. Before Bikini the world stood in awe of this new cosmic force. Since Bikini this feeling of awe has largely evaporated and has been supplanted by a sense of relief unrelated to the grim reality of the situation. Having lived with a nightmare for nearly a year, the average citizen is now only too glad to grasp at the flimsiest means that would enable him to regain his peace of mind.

It has been said that this psychological effect was intended from the start by those who arranged the Bikini experiment. But this view assumes a Machiavellian intelligence. The truth probably is that the American Navy, in its efforts to attract the attention of the public, desired to have a test of its own after having been defeated in its struggle with the Army for permission to co-operate in the development of the bomb.*

In reality the American public, after thirteen years of unrest and war, simply had neither the capacity nor the desire to be impressed by any further warnings or prophecies of terror. The people's growing apathy was due not only to the Cassandra-like utterances of the scientists but also to tranquilizing articles such as a report on Hiroshima by Major de Seversky, published in the *Reader's Digest,* which intentionally underplayed its hor-

* Although the first six thousand dollars provided from public funds for atomic research in the United States came from the Navy, General Groves would not even allow it to procure uranium for its own experiments. George Gamow, who during the war had worked out an early solution of the atom-bomb problem on behalf of the Navy, was unable to continue his studies simply because he had applied to the wrong branch of the Armed Forces for protection.

rors. If the citizens of any town in North Carolina, Kansas or Texas heard an atomic physicist say in the course of explanatory lectures that there was really no defense against the new bombs, a typical reaction noted by investigators from Cornell University who were trying to assess public opinion by questionnaires was, for example: "I am only one of many people who take life as it comes. If I have to live in a country where there are earthquakes, surely there would be no point in my going to bed every night in fear of an earthquake."

This new feeling of helplessness in the face of natural forces which man himself had the power to liberate was accompanied by a renunciation of civic responsibility. "I'm not worrying about it," said one of the average citizens interviewed in August 1946 by the Cornell investigators. "The government is sure to be taking precautions. Why should my heart be heavy over something I can't possibly control?"

Even the labor unions, which had originally attempted to mobilize their members against atomic armaments, became increasingly indifferent to the question. The following incident is evidence:

The members of a pacifist organization of workmen, led by their president, James Peck, decided in the summer of 1946 to demonstrate against the use of atomic energy for war purposes. The demonstration was to be staged outside the Oak Ridge factories, which were still working day and night at the production of explosive material for atom bombs. But the C.I.O. leaders acting for the factories stopped the demonstration, since any move towards abolishing atomic-armament factories might ultimately endanger the jobs of those workers.

To combat this public apathy a group of atomic scientists at the University of Chicago, headed by Hyman H. Goldsmith and Eugene Rabinowitch, founded a periodical, *The Bulletin of the Atomic Scientists*. Its object was to explain the social and political consequences of the new source of power. The idea first arose in discussions at a drugstore on 57th Street, opposite the campus. The editorial work was done in the basement of Eckart

Hall and the printing by contract with a small East Side newspaper for Czech immigrants. But the *Bulletin* exercised from the start an influence on leading American intellectuals far beyond its limited circulation. Nevertheless, the new journal, perhaps the most important publicizing venture to appear in the atomic age, was perpetually involved in the greatest financial difficulties. "To say that the *Bulletin* was founded on a shoestring would be to describe it as overdressed at birth," one of the editors recalls. "It lived for many months from hand to mouth, supported by the atomic scientists of Chicago, debts and Goldsmith's convictions."

In 1952, after years of struggle, the *Bulletin* appeared inevitably doomed to extinction. But at the last moment it was saved by the voluntary demise of the Emergency Committee for Atomic Scientists which the trend of the times had discouraged to the point of surrender. Its members, however, were able, before laying down their arms, to apply the last of their resources to giving the dying Chicago periodical the requisite "shot in the arm." The Emergency Committee of Atomic Scientists had been set up soon after the war at the suggestion of Albert Einstein to enlighten the public about what it might hope and fear from atomic power. The great scholar had been deeply shocked by what had followed his historic letter of August 1939. After Hiroshima he became the most resolute of those opposed to the atom bomb. Having traveled thousands of miles to escape nationalism and militarism in Germany, he had watched with horror the encroachment of these forces upon the American continent. But none of his passionately indignant speeches, manifestoes and protests seemed to have done any good. His anxiety for the world had eventually made him only too ready to sign repeated public petitions. Kowarski, one of Joliot-Curie's early associates, recollects that when he asked a group of American lecturers and students after the war what they were discussing so eagerly, he received the slightly ironical answer: "Oh, we're just wondering what we're going to say in Einstein's latest letter to the President!"

But by 1947 the greatest scientist of his generation had already recognized that all his efforts and those of his colleagues had failed to penetrate the public's stubborn indifference. In his discouragement he made the following statement to representatives of the foreign press: "The public, having been warned of the horrible nature of atomic warfare, has done nothing about it and to a large extent has dismissed the warning from its consciousness. It should not be forgotten that the atomic bombs were made in this country as a preventive measure. It was to head off its use by the Germans if they discovered it. We are in effect making the low standards of the enemy in the last war our own for the present."

During a walk with Ernst Straus, a young mathematician acting as his scientific assistant at Princeton, Einstein remarked, as though to console himself for his almost completely unsuccessful political efforts: "Yes, we now have to divide up our time like that, between politics and our equations. But to me our equations are far more important, for politics are only a matter of present concern. A mathematical equation stands forever."

It was not, however, the lack of understanding among many politicians, not the counteroffensive of the military authorities, nor even public apathy, that finally brought the scientists' crusade to a halt. It was above all the attitude of Soviet statesmen. It was evident from the way in which the subject was handled in the Russian papers that the Soviet Union, whether deliberately or because it knew no better, set less store by the importance of the atom bomb than did the West. The Russian newspapers said very little about the dropping of the bomb on Hiroshima and did not mention the Nagasaki bomb at all. In the first years after the end of the war almost nothing was done to inform the Russian people about the real nature and the danger of atomic warfare.

Nor, apparently, did the campaign of atomic scientists in the Western democracies for international control of atomic energy arouse any interest among their Soviet colleagues. The American atomic scientists made every effort, at the White

House and in the State Department, to ensure the participation of an eminent scientist in the Four Power Conference planned to take place in Moscow at the end of 1945. They gained their point at last, after some preliminary resistance by the statesmen. But when James Conant, the scientific spokesman, attended the Moscow meeting, he was given no opportunity whatever to bring up his carefully prepared suggestions for an exchange of scientists and international controls. Molotov sidetracked any discussion of such problems, postponing them to the next meeting of the United Nations. Conant returned home without having said a single word and expressed his well-grounded disappointment to his colleagues.

Nevertheless, the atomic scientists of the West hoped, almost without exception, that the Russians would at least show some serious interest in the American plan for the control of atomic armaments, which had been worked out with Oppenheimer's assistance and included many of their own ideas. The unmistakable rejection of the plan by the Soviet representative, Andrey Gromyko, on July 24, 1946, was a bitter blow for many of them.

Six days after Gromyko's speech to the Assembly of the United Nations President Truman signed the MacMahon Bill, establishing civil authority, for which the atomic scientists had fought so hard, over the "new power." But the victors were no longer much elated by a success which only affected domestic politics. They already suspected that in consequence of the tense international situation the soldiers would soon, after all, necessarily be calling the tune of atomic development. For it was they who would be the most important and favored customers of the Atomic Energy Commission controlled by the civilians.

As the debates in the United Nations Assembly on atomic-armaments control persistently hung fire, the Emergency Committee of Atomic Scientists tried to arrange a comprehensive exchange of views between Western and Eastern atomic scientists, hoping that the outcome would enable them to show their respective statesmen a way out of the impasse. Such an international conference of nuclear physicists was planned not only to clear up misunderstandings between Communist and non-Communist countries, but also to extinguish the competition between Western laboratories, due to the official policy of secrecy to which they were bound by their governments. The British, for example, had some right to complain that the Americans had not adhered to the agreement concluded in 1943 at the Quebec Conference about the mutual exchange of atomic data. Even official American publications dealing with the history of the production of the atomic bomb, the British considered, minimized the contributions of British, French and Canadian scientists. It seems typical of this attitude that in the official American film of the Bikini test the voice of the British scientist Ernest Titterton on the loud-speaker system, counting the seconds that elapsed before the explosion, was cut out and replaced by a voice with an American accent.

At the great international family reunion of physicists contemplated, the misunderstandings that had arisen during the war would have been cleared up and an attempt made to revive the wide international brotherhood of science.

Harrison Brown, in the name of American scientists, submitted to Andrey Gromyko, through a Polish delegate to the United Nations, this proposal to call an international conference of this kind to enable scientists to discuss questions relating to the control of atomic armaments. Jamaica was suggested as the place of meeting. Brown states:

The secret police of both groups of powers, we thought, might be allowed, before the meeting, to install their listening apparatus in the large hotel where it was to take place. They might perhaps also make use of that opportunity to get to know one another a little. An answer from Moscow was received in a surprisingly short time. We were invited to the New York quarters of the Soviet delegation to the United Nations, on Park Avenue, in order to hear it. Our hearts beat high with hope. But when we arrived we found Gromyko in quite a different mood from that of our first meeting. He confined himself to reading out to us—and to the microphones which were probably concealed in the room—an extremely formal rejection of our proposal.

Only on one occasion did a friendly voice reach the Western atomic scientists from the huge prison in the East. Referring to the explosion of the experimental bombs at Bikini, Kapitza called upon his Western colleagues to go on fighting to prevent the use of atomic energy in war. The former assistant and friend of Rutherford declared: "To speak about atomic energy in terms of the atomic bomb is comparable with speaking about electricity in terms of the electric chair."

That was the last time that the West heard from Kapitza. It was not until ten years later that some American scientists, who had been invited to attend a large gathering of physicists in the Soviet Union, learned what had become of him. Soon after the statement he had made in July 1946, which had been directed, it now appeared, not only to foreign but also to Soviet atomic scientists, Stalin had placed him under house arrest because of his refusal to work on the production of Soviet atomic weapons. At the same time he had been removed from his post as Director of the Institute for Physical Problems which had been established for him. He remained practically confined to his home at Zvenigorod for seven years. During the whole of this time erroneous reports were continually appearing in the Western press that he and no other was responsible for the Soviet atom bombs. The American physicists who visited Moscow in the summer of 1956 were told that a number of other Russian nuclear experts like Kapitza had been sentenced under Stalin to deportation and compulsory labor for having refused to work on atom bombs. But it was only a small minority which dared such a resistance.

By the spring of 1947 it had become clear to everyone that the scientists' crusade had failed. The atomic-armaments race was in full swing. The new scientists' organizations had been definitely thrown on the defensive. The return journey to the armaments laboratories had begun.*

General Groves had been right. He humorously remarked later about the rebels: "What happened is what I expected, that after they had this extreme freedom for about six months their feet began to itch, and as you know, almost every one of them has come back into government research, because it was just too exciting."

Actually General Groves simplified the true situation. Only a minority of American atomic scientists were perfectly free agents in deciding to resume participation in government-sponsored research. Most were compelled to take this step, because they would have had no choice, otherwise, but to change their profession. They could not help noticing that while they had been engaged in promoting civil control of atomic energy, the military men had cleverly turned their flank by infiltration into the strongholds of the scientists themselves, the universities.

The universities during the war had found a new and extremely wealthy patron in the Armed Forces. Although they were bound to regard these funds for weapons research as merely temporary until the war was over, the scientists had greatly expanded their departments of physics, chemistry, technology and biology because of military financial support. After the war university presidents, occupied with their peacetime budgets, were visited by representatives of the Office of Naval Research or G6 (for research) of the War Department. They explained: "We are ready to go on financing you. There is no need for you to close any of your expanded laboratories or dismiss any of your staff. We shan't even ask you to work

* A questionnaire—not intended for publication—by the Federation of American Scientists, which had been in the forefront of the struggle to prevent the use of atomic energy in warfare, registered 243 affirmative votes in the spring of 1947 in answer to the question: "Do you think that the United States should proceed with the production of atom bombs?" Only 174 negative replies were registered. At Los Alamos there were actually 137 votes cast against the discontinuance of bomb production and only 31 in favor of it.

on inventions which we can use immediately. You may devote yourselves to theory. We want to promote a flourishing school of research. In this century the strength of a nation is measured not only by reference to its arsenals but also by its laboratories. Go quietly ahead with your peacetime tasks."*

Thus by the end of 1946 the Armed Forces had already spent many millions of dollars financing not only their own research organizations but also the university laboratories. As early as the end of October 1946 Philip Morrison indicated his anxiety about the situation during the annual forum on public affairs conducted by the New York *Herald Tribune:*

At the last Berkeley meeting of the American Physical Society just half the delivered papers . . . were supported in whole or in part by one of the services . . . some schools derive ninety per cent of their research support from Navy funds . . . the Navy contracts are catholic. They are written for all kinds of work . . . some of the apprehension that workers in science feel about this war-born inflation comes from their fear of its collapse. They fear these things: the backers—Army and Navy—will go along for a while. Results, in the shape of new and fearful weapons, will not justify the expenses and their own funds will begin to dwindle. The now amicable contracts will tighten up and the fine print will start to contain talk about results and specific weapon problems. And science itself will have been bought by war on the installment plan.

The physicist knows the situation is a wrong and dangerous one. He is impelled to go along because he really needs the money. It is not only that the war has taught him how a well-supported effort can greatly increase his effectiveness, but also that his field is no longer encompassed by what is possible for small groups of men. There is a real need for large machines—the nuclear chain reactors and the many cyclo-, synchro- and beta-trons—to do the work of the future. He needs support beyond the capabilities of the university. If the

* In order to prevent such a development the scientists had suggested the establishment of a National Science Foundation, to be financed by the government, which would provide the universities with funds for theoretical research on problems of public interest. But this foundation, largely because of disagreement among the scientists themselves, came into being only some years later. Nor did its annual budget represent more than an insignificant percentage of the funds provided by the military authorities.

O.N.R. or the new Army equivalent, G6, comes with a nice contract, he would be more than human to refuse.

The situation foreseen by Morrison came about more swiftly and completely than the most pessimistic observer could have anticipated.* In the universities, once the homes of free speech throughout the world, the spirit of secrecy took possession. Some of their research was under military safeguards and law. Invisible barriers and trenches were placed around them. Professors began to have secrets and could only talk to one another, like priests of some peculiar religion, in a special language, when they wished to discuss their affairs. Since but a few people knew what they were really doing, even those who could only with difficulty have reconciled their work with their consciences, remained at their posts. For when secrecy about armaments governs, one need fear no other censure but the military.†

* A decade later *Business Week* for January 12, 1957, reported under the heading "Defense Department, Leading Patron of the Sciences": "In the United States military expenditures for scientific research and development have jumped from an annual average of 245 million during World War II to 1.5 billion this year. This trend will continue to rise. . . . Indirect military and development expenses . . . [reach] . . . at least 3.6 billion."

† The following amusing parody of the new jargon used by armaments scientists circulated at that time in the University of Chicago: "Gentlemen, I feel that to set you right I will have to tell you the true story of the X Plant operations. Of course, I must ask you to keep it strictly secret, because I am the only one who knows. So please, if you do tell anybody what I am about to tell you, be sure to ask them not to tell.

"Here is the whole procedure. They are taking plumscrate, raw plumscrate mind you, and putting it into ballisportle tanks. These are called ballisportle tanks merely because the inside is coated with quadrelstitle thus preserving the full strength of the plumscrate.

"Next, this is taken to the sarraputing room, where only expert sarraputers are employed. Here, of course, is where they add thungborium, the ingredient which causes the entire masterfuge to knoxify. After this jells into five pound ignots a heavy coating of spurndaggle is applied, causing the entire product to disappear. The workmen, known as spurndagglers, also disappear.

"Now the invisible compound is taken to the abblesnurting building, also invisible, where glass snaggle hooks are applied for carrying.

"This completes the operation. Delivery is the next problem. At twenty minutes past twelve on the third Tuesday nite of each month eight hun-

On March 21, 1947, President Truman issued the "loyalty order," calling for a thoroughgoing police investigation of the political and moral reliability of all government officials. Since most of nuclear research both in and out of the laboratories was directly or indirectly financed by the Federal government, the atomic physicists were particularly involved in the operation of the decree. Some idea of the atmosphere which prevailed in the "atom cities" may be obtained from the following story told to a scientific congress by Dr. Swartout, Director of the Radiochemistry Department at the Oak Ridge atomic-research laboratory:

> One evening in the summer of 1947 a scientist—and by the term "a scientist" I indicate only a typical case—was aroused from his dinner by a knock on the door. At his door stood a uniformed guard, who demanded that the man turn over his badge, his means of access to the town in which he lived and the installation at which he worked. Because the guard could give no reason, the man called his supervisor for an explanation, who turned out to be completely unaware of what was going on. After calls to successively higher authorities the man was told to comply with the guard's orders and to report to the installation director's office early the next morning. Confronted by officials on the next day he was told that the F.B.I. investigation had disclosed information which forced the Commission to consider him to be a questionable security risk; that he would be permitted to submit a statement in his defense regarding his character, loyalty and associations; and that this would be reviewed by an A.E.C. board in Washington. In the meantime he would get a temporary pass to admit him to his home but not to his place of work.
>
> Imagine yourself in his position. If you were asked to defend your character, loyalty and associations, what would you do? Against what! Whom had he known, what had he done or said which would bring this accusation against him!

---

dred men (800) known as shizzlefrinks because their brains have been temporarily syphoned from their heads are lined up in single file, each given two ignots of the Oustenstufftingle (that's the name of the finished product) and away they march over the hills to Blank Blank, where they trade the finished product for enough raw materials to make another batch of Oustenstufftingle."

The story—and here it was far less typical—had a happy ending. The accused was reinstated in his position. "But," Dr. Swartout goes on, "all of this required time, several months, during which he endured the suspense of not knowing whether his job and scientific career would be erased and during which he was not permitted to work. In the case of direct employees of the A.E.C. this was also a period without an income. Employees of contractors fared better, but a vacation of several months under such constant mental stress is not advised for rest."

There were hundreds of such cases in these "bitter years." Statistics alone cannot describe them, for no figures can comprise the whole burden of anxiety, fear and grief borne by all who were involved, because of some unsupported denunciation or some long-forgotten episode in the past. No later revision on the grounds that the anonymous accusers and the judges (who as a rule tried these cases reluctantly) had violated basic civic and human rights, could compensate the victims of such crude proceedings for all they had suffered. Though they were never brought before a regular court of law, they lived under suspicion of having betrayed their country. They were spied on by the government, mistrusted and avoided by most of their neighbors. Many of their colleagues no longer dared to talk to them. It was a time of sentences of banishment, of exile endured in their own country, a time of grief and shame that drove men to suicide.

From 1947 on, the atmosphere in which the Western scientists lived became more and more oppressive. The new methods used by Washington, the center of political power in the West, influenced the mental climate of London and Paris. Soon even in England and France unpopular scientists were being examined by loyalty committees, deprived of their passports and dismissed from their posts. Friendships between men of science broke down under the strain of mistrust and fear. Scientific correspondence that had lasted for decades came to an end. Even in the laboratories of the Western world people started whispering to one another, anxiously on the watch for the state's long ears, as had been true before only in totalitarian countries.

Yet even such drastically restricted freedom, with its unhealthy climate of suspicion, accusations and time-wasting defense against false charges, was still preferable to the total slavery which reports from behind the Iron Curtain said was the fate of scientists in the Communist states. The persecution, in particular, of the Soviet geneticists who had been disciplined or even, like the famous Vavilov, murdered, for their deviations from the theories of Lysenko, aroused deep sympathy in the West.

During the "beautiful years" atomic scientists had been short of funds for their research. But in compensation they had been able to work in a free and fortunate atmosphere. Their studies had gone almost unnoticed, yet for that very reason they themselves had been all the more respected. There had only been a few of them and they had known one another across great distances of land and sea. Now there were perhaps a hundred times as many. Their science had become the fashion. Their congresses were like mass meetings. Many people feared or even hated them. They were regarded as important personalities—so important indeed that on occasion they were not even allowed to die alone.

In those days a desperately sick man was once brought unconscious, under strict military guard, to the Letterman Hospital in San Francisco. He was taken to an isolation ward. At first an armed sentry was posted at the door of the ward. Later the soldier was withdrawn, as his presence had aroused comment in the hospital. All the doctors and nurses in attendance had been previously tested for political reliability. It was impressed upon them that they must immediately forget anything the man might mutter in his delirium.

William G. Twitchell, the patient in question, was a nuclear chemist from Minnesota, thirty-six years old. He had for some years occupied a post of some authority in the radiation laboratory of the University of California. As one branch of this world-renowned research establishment was at that time working exclusively upon the improvement of atomic weapons, Twitchell probably knew some important atomic secrets. The circumstances in which the young man had fallen ill were never made public. At any rate Fiedler, Director of the Berkeley Security Division—the same office in which Oppenheimer had made his first confession in 1943—did his best to hush up the affair. He had requested the Army to transfer the patient to a hospital of its own, where stricter security regulations could be enforced than at a civil hospital.

The case came by chance, six months later, to the attention of a correspondent of the New York *Times*. But even he was unable to discover the precise nature of Twitchell's illness. The scientist may simply have collapsed under the strain of the duty of silence imposed upon him. This happened to a naval officer during the war, who was serving in the atomic laboratory at Oak Ridge. He was arrested by security officials in the act of telling people in a crowded railway carriage about the work being done in the atomic city. A special small clinic, with doctors and staff in attendance, was established to deal with this one man who had become mentally unstable—it was not considered advisable to transfer him to any private, much less public, mental hospital.

No such arrangements were found necessary for Twitchell, who died a few days after his admission. Not one of his friends or relatives was allowed to be with him during his last hours.

Sixteen

"Joe I" and "Super" *(1949-1950)*

Toward the end of August 1949 a United States Air Force "flying laboratory," installed in a B-29 to assemble data for the protection of bomber crews, made a disturbing discovery. The photographs brought back from a flight somewhere in the Far East showed clear, unaccountable traces of radioactive matter in the atmosphere. In addition to the usual threadlike white tracks which particles from outer space commonly made on the negative, many other new lines had become perceptible. The phenomenon seemed so unusual that it was reported immediately, in code, to Washington. At once orders were issued for RD aircraft, specially equipped for radiation detection, to investigate. They took samples of rain from high clouds and also, using a kind of flypaper, certain microscopic particles of ash from the highest layers of the atmosphere. These were subjected to thorough radio-chemical analysis. Not until then did the scientists of the Air Force and the Atomic Energy Commission venture to put into words what they had suspected from the beginning. The radioactivity discovered must have originated from an atomic explosion somewhere in Soviet Asia.

Among the few experts who learned this there was tremendous surprise. People had grown used to believing the prophecies that the Russians would possess atom bombs, if ever, not before 1956 or 1960. Air Force technologists, who had suggested an earlier date, 1952, were considered by the Army and Navy to have been exaggerating. But it now looked as though even they had been overoptimistic.

The strategists of the Pentagon, after recovering from their original shock, began to console themselves by conjecture. Probably, they thought, the heightened degree of radioactivity had not arisen from the test of a bomb ready for use in warfare but from some negligence or clumsiness which had caused an accidental explosion in one of the Russian atomic laboratories. And yet an explosion on so great a scale must indicate that the Soviet Union already disposed of considerable quantities of fissile material. How had the Russians contrived to produce so much U235 or PU239 (plutonium)? Could they possibly have built the comprehensive installations necessary for the purpose in the four years since 1945? Even to this fateful question an answer was found which minimized the importance of the event. It was supposed that the fissile material had not been produced behind the Iron Curtain but secretly smuggled into the Soviet Union by spies. This did not sound very credible, for during the immediately preceding months Senator Hickenlooper had subjected all the operations of the Atomic Energy Commission to a painstakingly detailed public examination. In this process only four grams of U235, in all, had been reported missing.

The widespread underestimation in the West, during the first four years after the war, of Russia's capacity to construct atom bombs within a foreseeable time is almost more astonishing than the earlier overestimation of Germany's atomic potential. Until the end of 1945 the Russians had mentioned quite openly in their technical writings and even in their daily press their great interest in nuclear physics and their studies in that field.

Two institutes in Leningrad, the Radium Institute and the Institute for Technical Physics, and two in Moscow, the Lebedev Institute and the Institute for Physical Problems, as well as an institute in Kharkov, had been concerned to some extent, ever since the beginning of the nineteen-twenties, with nuclear research. That the Russians possessed large deposits of uranium ore—and knew it—was clear from the publications of the well-known geologist Vernadsky. He and his pupils had begun by 1921, on Lenin's instructions, the exploration and description of all deposits of raw materials throughout the Soviet Union.

As soon as the news of Otto Hahn's discovery was published, Soviet scientists had begun to study its significance and estimate its possibilities with as much enthusiasm as their colleagues in the West.* An official and public congress dealing with problems of nuclear physics took place in Moscow in 1939. In April 1940 the Soviet Academy of Sciences announced in its monthly bulletin the formation of a special Uranium Problem Commission. All the leading Russian physicists belonged to this Soviet Uranium Society, including Flerov and Petrzak, who had been the first to discover the spontaneous fission of uranium when in 1940 they carried out certain experiments in a shaft of the Moscow subway.

As early as 1939 A. I. Brodsky published an article on the separation of the uranium isotopes, while Kurschatov and Frenkel, at about the same time as Frisch, Bohr and Wheeler, gave theoretical explanations of the fission processes in uranium. In the 1940 New Year's Eve issue of *Izvestia* an article entitled "Uranium 235" contained the following passage: "Mankind will acquire a new source of energy surpassing a million times everything that has hitherto been known . . . we shall have a fuel which will be a substitute for our depleting supplies of coal and oil and thus rescue industry from a fuel famine . . . human might is entering a new era . . . man will be able to acquire any quantity of energy he pleases and apply it to any ends he chooses." In October 1941 Kapitza stated in a lecture published in many Soviet newspapers that "Theoretical calculations prove that . . . an atom bomb . . . can

* From the start, uranium fission interested not only Russian men of science but also the Russian government. When Kaftanov, the Soviet Minister for Education, visited Berlin in 1939, he was particularly anxious to see Hahn's laboratory and interview him personally on the subject of his experiments. His request was granted.

easily destroy a large city with several millions of inhabitants."

In 1941, after the German invasion, the Russians seem to have abandoned their program of atomic research for the time being. The Rand Corporation, which acts under the orders of the United States Air Force and, among other activities, issues reports on technical progress in the Soviet Union, published in 1956 a study with the following statement: "But the Russians apparently dismissed the idea that it [the bomb] would be feasible for the war then raging. They made no attempt to conceal the fact that they had stopped atomic research; and they apparently did not assign priority to the subject of atomic energy in their foreign espionage. . . . By 1943 the Russians had resumed an atomic development program with the apparent intent of trying to acquire nuclear weapons."

The earlier false conclusions drawn in America from the failure of the atomic project in the Third Reich also contributed to the underestimation of Russian atomic research and the progress made by the totalitarian Soviet state. Under Stalin, as under Hitler, officialdom had attacked modern physics on ideological grounds. The quantum and relativity theories, and "Einsteinism"—as it was called in the Russian technical press—were condemned as "idealist" and "reactionary." But this was as far as the resemblance between the two dictatorships went. In the National Socialist Third Reich natural science was not encouraged. But in Stalinist Russia it was given every possible material support. Not only was the profession of physicist among those which had the highest salaries and standards of living, but their institutes received exceptionally large funds for their projects. The Russian nuclear experts were therefore able, before 1939, to build the first cyclotron in Europe. By 1941 they were erecting two more of these gigantic atom-smashing machines. One was designed to attain a radiation strength three times as great as that of the biggest apparatus then operated in the United States.

Ruggles and Kramish, two of the specialists studying Soviet atomic research for the Air Force, conclude that "far from starting a nuclear research program from scratch in 1945 the

Russians should not have been by that date too far behind the knowledge and skill that had been achieved in the United States. These findings might well occasion some surprise that it took Soviet industry four years more to produce the atomic bomb exploded in 1949."

This realistic assessment of Soviet atomic development was not made until 1956. During the first years after the war it would probably have been dismissed in the United States as exaggeration, like Molotov's declaration of 1947 that the atom had no more secrets for Soviet scientists.

In the anxious days that followed the discovery of the first Soviet atomic explosion of August 1949, the Washington authorities were fortunately not content with deceptive consolations of conjecture. They called in a committee of specialists to make a further statement, based on all the available evidence. The committee held several sessions under the chairmanship of Vannevar Bush, with the participation of Oppenheimer and Bacher. After examination of all the information they not only came to the conclusion that an atomic bomb must have been involved but were also able to supply data about its probable composition and the force of the explosion. The American scientists were by that time so certain of the existence of the Soviet atom bomb that they gave it a name, "Joe I," in "honor" of Joseph Stalin.

It was then necessary to apprize President Truman and the Congressional Joint Committee on Atomic Energy of the appearance of Joe I on the scene. Both the President and a leading Republican Senator, Vandenberg, reacted to the information with the same question, which revealed the depth of their consternation: "Where do we go from here?" The first decision necessary was whether to publish this secret story to the world. Secretary of Defense Johnson was in favor of withholding it, fearing it might cause panic in America. Johnson was outvoted. On September 23, 1949, President Truman read out his brief and very carefully phrased message, stating that an atom-bomb explosion had taken place in the Soviet Union.

Even this information did not succeed in shocking the great masses of the people out of their helpless indifference to the atomic peril, but the excitement among American atomic scientists increased. Almost without exception they had pointed out, since 1945, that the United States monopoly of this weapon could only be brief. They now believed that there was practically no hope of ending the atomic-armaments race, that, on the contrary, it was likely to become more acute. Their anxiety found visible expression in a symbolic act. The cover of the *Bulletin of Atomic Scientists* bore every month a design representing a minute hand pointing to eight minutes to twelve. The hand was now advanced to indicate three minutes to the hour. The end of time had come nearer still.

In the discussions among those "in the know" that arose after the news of the explosion in the Soviet Union one word was repeated continually which outsiders would scarcely have understood. It was "Super." The *Bulletin,* which had developed into an internationally respected organ and forum of debate among atomic scientists, had for years of its own accord suppressed any mention of what the term meant. It was considered the best policy not to draw anyone's attention to the monstrous further growth in armaments technique at which the word hinted.

For Super was a bomb which might well be a thousand times as powerful as that which had razed Hiroshima to the ground. Unlike the ordinary atomic bomb it was an "open-ended weapon," of unrestricted range.

Such a bomb could be constructed only if the powerful natural processes taking place in the interior of the sun were successfully reproduced on earth. Quantities of energy were perpetually being released in that flaming heavenly body by the fusion of hydrogen atoms. The forces liberated were incomparably more powerful than those let loose in uranium fission.

The Super first became the subject of investigation as far back as the summer of 1942. At that time Oppenheimer had gathered about him at Berkeley a small group of theoretical physicists to consider the question of the best type of atom bomb to make. During the discussions Teller, who had been engaged for some years, at Gamow's suggestion, on the study of such thermonuclear reactions in the stars, had indicated the possibility of a fusion of this kind, the logical next step after the fission bomb.

At the University of California at Berkeley, most of the University's undergraduates were already on vacation or away on military service. The scientists—seldom more than seven—participating in the discussions had practically the whole campus to themselves. It was there, on a green lawn among high cedars, or in one of the many-windowed lecture rooms, to the accompaniment of the regular chiming of the hours from the campanile, like music of the spheres, that for the first time the conversation turned to the idea of man-made suns. Those were days, as Teller later recalled, filled with "a spirit of spontaneous expression, adventure and surprise." The deep thrill at the discovery of the new dimensions of human knowledge and power made most of them forget that they had really met to design an instrument of death.

As a result of the Berkeley conversations, the decision was made to begin by concentrating mainly on the construction of the uranium bomb, but also meanwhile to devote further serious attention to the problems of the Super bomb. One of the questions to be considered was peculiarly sinister. The possibility had been mentioned at Berkeley that once the thermonuclear processes had been set in motion by the explosion of a bomb, they might affect the atmosphere and the waters of the earth. An irresistible global chain reaction might be released by the Super, which would transform the entire planet in a short time into a flaming and dying star. The study of this monstrous idea was at first assigned to two theoretical physicists, Emil Konopinsky and Cloyd Marvin, Jr. They both returned a reassuring answer to the question, but not everyone was convinced by it. An appeal for a final decision was made to Gregory Breit, a physicist celebrated for his sagacity and precision of thought.

Gregory Breit had been brought to the United States as a fifteen-year-old boy, in flight from the pogroms of tsarist Russia. In America he had found a quiet spot to live in, suitable to his retiring nature. He was able to read, think and teach without having to worry much about the rest of the world. All that was changed one fine day in 1940. Professor Breit was taking one of his habitual walks in a Washington park when a car drew up beside him. He was asked if he would like a lift. In ordinary circumstances he would probably have declined, but that day he felt rather tired and gratefully accepted the invitation. It soon turned out that the good-natured driver was a member of the Scientific Research Department of the Navy. He was attracted by the shy professor, and asked him to call at his office in a day or two. He told Breit that the Navy had a specially interesting problem in physics to solve.

The Professor agreed to come. He had little inclination to work for the promotion of war and destruction, but the naval officers did not ask him to do so. They were looking for a man who could suggest some means of protecting their warships against the new German magnetic mines; it was only a matter of preserving men's lives and property. Breit consented to help. He began to work for the Navy and his ideas soon put the physicists of the Research Department on the right path.

Not long after, Breit was again summoned to a government department. He was told that he was the only man capable of co-ordinating and directing work on a new bomb. The officials added that there was no intention of using the new bomb in warfare. It was merely to serve as a deterrent in case the Germans developed a similar weapon. Breit's work would be contributing to the preservation of the whole nation from a catastrophe. "I'm a bad administrator," Breit objected. "You could not have hit upon anyone less suitable for the work of coordination." But the others argued: "We haven't anyone else who is free to undertake the job and an American citizen.

Nearly all the other physicists engaged in this affair are foreigners."

The peace-loving Professor was thus persuaded to preside over the first committee to study "fast fission" in Washington—such was the phrase used to describe the uncontrolled chain reaction which takes place in the atom bomb. After a few months Breit, to his great relief, was permitted to resign this responsible post. He assumed that he would now be able to return to his own scientific work.

But he was soon called upon to advise on another problem, the "global chain reaction." The first time he had been asked to help to prevent the destruction of warships. Next, it had been a case of the possible destruction of the United States. Now the possibility was the destruction of the whole world!

The entire responsibility would be his alone; his judgment was to be accepted as final. As the matter was strictly secret he could not count on other physicists being simultaneously made aware of the problem. Suppose he gave the wrong answer to so important a question, one which had never before been put to anyone, even in myth or legend? Supposing he overlooked some factor in it? Supposing he said, "All right, the risk you mention, so far as can be foreseen by the human mind, does not exist," and then it turned out that he was mistaken? Had not the possibility of release of the forces that slumbered within the atom long been discounted by the most famous of scientists? Was not another such error of judgment still conceivable?

It would have been understandable enough if Breit had declined to undertake the task demanded of him, to assume the superhuman responsibility; but he had to remember that in such a case the task would certainly have been entrusted to another scientist who might be less judicious than himself. Also, he could rely absolutely on his own conscientiousness.

For considerable time, during which the whole burden of the fate of the earth and its inhabitants rested upon the Professor's narrow shoulders, he calculated and meditated day and night. At last he could do no more than work out his calculations and submit them to his taskmasters. He now believed that he had proved beyond human doubt that an unprecedented encroachment upon the light elements of the earth of the reactions liberated in a thermonuclear bomb could not occur in any circumstances, that this was contradictory to the fundamental laws of Nature.

And yet other doubts must then have tormented Breit. His opinion had indeed eliminated the greatest, probably the greatest conceivable, obstacle to the future construction of the Super bomb. But wouldn't he now share the responsibility if such bombs were ever used, not experimentally, but wantonly and deliberately, to bring destruction on a planetary scale upon the world?

When the lovable little man reached this point in his reflections he must have suffered terrible mental distress. He had never done anything, throughout this frightful war, but try to help avoid a worse catastrophe. What could one do that would not lead to guilt?

At the discussions in Berkeley it had not been supposed that the Super bomb would take very long to construct. During the laboratory experiments between 1943 and 1945, however, the goal appeared to recede further and further. After all, the ordinary atom bomb had first to be developed. It was an indispensable preliminary to the Super. For only a uranium-fission bomb of this kind, built into the hydrogen bomb as a fuse, could generate the enormously high temperatures required to trigger thermonuclear reactions. This task turned out to be more difficult and protracted than had been anticipated.

Much to Edward Teller's annoyance, the Super project was more and more definitely shelved as time went on. Even he himself was not at first allowed to work on it; there were more urgent things for him to do. But Teller was not made to march with the rank and file. Systematic work simply did not interest him. Serious friction resulted. His chief, Hans Bethe, reported later:

> I relied and I hoped to rely very heavily on him to help our work in theoretical physics. It turned out that he did not want to co-operate.

He did not want to work on the line of research that everybody else in the laboratory had agreed to as the fruitful line. He always suggested new things, new deviations. He did not do the work which he and his group were supposed to do in the framework of the theoretical division; so that in the end there was no choice but to relieve him of any work in the general line of the development of Los Alamos, and to permit him to pursue his own ideas, entirely unrelated to the World War II work, with his own group outside of the theoretical division.

This was quite a blow to us because there were very few qualified men who could carry on that work.

The gap left by Teller at that time was filled by Rudolf Peierls and Klaus Fuchs. Teller proceeded, with a small group, to work on the problem of the Super, which he called "my baby."

In a community as closely knit as Los Alamos during the years of war an outsider like Teller was bound to attract special attention. In time he excited envy, irritation and even hatred. Other scientists submitted to military discipline, though they would formerly never have dreamed of such a thing. Punctually at an early hour every morning they disappeared behind the barbed-wire fences of the Technical Region. Teller got up late, worked at home and then went for long, lonely walks. In a university town these habits would not have aroused much comment. But on the Hill questions were asked, such as: "What's he really doing here? Why doesn't he have to obey the same regulations as everybody else?"

Complaints about Teller were laid before Oppenheimer, the director of the laboratory. Many were petty fault-finding. Surely the Tellers, with only one child, had one room too many? Why should they put their little son's playpen right in front of the apartments, where the cycle racks ought to be? Was Teller to be allowed to play the piano late at night, disturbing all the neighbors?

Oppenheimer took little or no notice of this tale-bearing. He had been told that Teller criticized him severely but also admired him. In many ways the two men were alike. They were both spurred on by a similar burning ambition. They both felt themselves to be immeasurably superior to their fellow men. They were both, as Bethe, who worked with them for many years, remarked, "almost more like artists than scientists."*

The oversensitive Oppie was perfectly well aware that all was not right in the relations between himself and this unusual colleague. He knew that in spite of their frequent meetings they had never really made contact with each other. For that reason he was particularly careful never to take any step which Teller might construe as indicating animosity.

On the other hand Oppenheimer never praised Teller as much as he probably expected. A witness remarks: "If Oppie had only occasionally, for once, said a few good words for Edward in those days, such as he very well understood how to address to almost any mechanic, the destinies of both men might perhaps have been different." This observation dates from a much later time, when the lack of sympathy between Oppenheimer and Teller had grown into a serious dispute with important consequences.

At the end of the war Teller did not at first join the general return to the university laboratories. During the war evil tongues had hinted that he envied Oppenheimer his post as director. It was now said that he considered himself to be a suitable successor, though no one but himself, it appeared, could imagine Teller as a satisfactory administrator.

Oppenheimer's actual successor, Norris Bradbury, may have heard something of these rumors. He sent for Teller and immediately offered him the second most important post in the laboratory, that of head of the Theoretical Division, left vacant by Bethe's departure.

* A former close collaborator, in a letter to the author, gave a notable description of Teller's character in the following terms: "I got to know him rather better while he was helping me with a chapter of my . . . book. He is a typical modern thinking machine, by no means without a heart or lacking sensitivity. But these last two faculties in him are on a very ordinary level and quite incapable of competing with the vigor of his intellectual pleasures."

A conversation took place between the two men which was full of unfriendly undertones. Teller declared in his usual aggressive style: "Let us see if we could test something like twelve fission weapons per year or if, instead, we could go into a thorough investigation of the thermonuclear question." Bradbury answered: "That is, as you must know yourself, unfortunately out of the question." Thereupon Teller refused Bradbury's invitation to remain permanently at Los Alamos and went off to the University of Chicago.

During 1946, however, Teller did return for a few days to Los Alamos to attend a special conference. The subject of the discussions, to which about thirty physicists had been summoned, was the Super. The majority of those present concluded that the development of such a weapon was bound to be protracted and complicated. A minority, led by Teller, maintained, on the contrary, that the bomb could be constructed in two years. The meeting then broke up. One of the participants in particular must have been deeply impressed by Dr. Teller's arguments; he did not hesitate to convey the fact to his contacts. This man was Klaus Fuchs, whose last important information given to the Russians dealt with that final conference on the Super.

In his chair of physics at Chicago Teller continued to advocate construction of the Super bomb. He demanded, for example, that the Emergency Committee of Atomic Scientists should not only listen to him but itself call for the construction of this terrible weapon, a demand which aroused great indignation in Einstein, the Chairman of the Committee. He sternly refused to comply. To Teller his attitude seemed illogical. The world situation after 1947, particularly since the Communist *coup d'état* in Czechoslovakia in February 1948, reminded Teller of the period between 1939 and 1941 when he had belonged to the small group around Szilard agitating for the construction of the uranium bomb. Was the position now so very different? he asked. There was again a danger that a totalitarian state might threaten freedom by a weapon to which there was no answer but "counterterror" by the same weapon. Why should Stalin be trusted more than Hitler?

But oddly enough Teller also took every opportunity to advance the cause of world government; it seemed to him to be the only hope of preserving peace. He used to say to other atomic scientists: "It won't be until the bombs get so big that they can annihilate everything that people will really become terrified and begin to take a reasonable line in politics. Those who oppose the hydrogen bomb are behaving like ostriches if they think they are going to promote peace in that way."

But Teller did not go so far as Harold Urey, who was also in favor of a world government. Urey, after vainly acting as a front-line champion of international control of atomic armaments, was now induced to call, actually, for a preventive war, so that humanity could enjoy peace and freedom again when it was finally over.

Until the explosion of Joe I became known, Teller's campaign for the construction of the hydrogen bomb found little support. The moment the news broke, all those who had formerly nicknamed him, rather derisively, the "apostle of the Super," remembered his warnings. People who were convinced that the armaments race was inevitable considered that Joe I would have to be trumped with the Super if the lead in the atomic field were to be retained.

Or was it already too late for that? Might not the Russians already actually be ahead in this sinister contest of speed? Such was the question Luis Alvarez asked himself. After the conclusion of his mission in Tinian he had returned once more to theoretical research at the Berkeley Radiation Laboratory. An entry in his diary reads:

> October 5, 1949. Latimer and I independently thought that the Russians could be working hard on the Super and might get there ahead of us. The only thing to do seems to get there first—but hope that it will turn out to be impossible.

Alvarez at once consulted Ernest O. Lawrence, who had been thinking along the same lines. They determined to get in touch with Teller immediately, but they did not know where he was and a telephone call to his apartment in Chicago remained unanswered. Teller, restless as ever, had obtained a year's leave from the University, to enable him to resume work in Los Alamos for a while. He had begun by going abroad for a few weeks. The news of the Soviet bomb reached him, as it did the general public, on September 23, 1949, as he was passing through Washington. He immediately phoned Oppenheimer to hear how he was taking the announcement, but Oppenheimer did not seem to be worried. He replied simply: "Keep your shirt on!"

Teller hurried with all speed to Los Alamos. There, on October 6, Alvarez and Lawrence at last contacted him by telephone, but the line was bad. The two physicists at Berkeley decided that, since they were to fly to Washington in two days' time, they would break the journey at Los Alamos to have a detailed discussion with Teller.

Los Alamos now had nearly ten thousand inhabitants. Amazing changes had taken place in the town since its decline in 1946. In addition to the intensive resumption and extension of the armament program considerable sums had been voted to build more laboratories and houses. There were now well-paved streets, a Community Center with a meeting hall, a movie theater and all kinds of shops. A large hospital had arisen, an excellent town library, good schools and a thriving sports club, called the "Los Alamos Atomic Bombers." A stadium was being built, named after Louis Slotin, the young atomic expert now revered as a martyr to the bomb.

Alvarez and Lawrence took an air taxi from Albuquerque up to the Hill. Teller carried them off to his house in the "Western Area," where the senior scientists now occupied cozy little villas. Their party was joined later by Gamow, who had recently arrived in Los Alamos as a temporary consultant, and by the gifted Polish mathematician Stan Ulam.

Ulam, during the years 1946 and 1947, and J. L. Tuck, an Englishman, had composed some extraordinarily interesting studies on thermonuclear problems. These included the effects produced by converging shock waves from hollow charges. An enormous temperature is produced, sufficient for fusion. But at that time, during the first years after the war, calculations of thermonuclear reactions had been held up chiefly because the calculating machines then in use could not cope with the problems that arose.

In the course of the discussion among these five men at Los Alamos the question was asked whether the supposed Russian interest in the new type of bomb had not been overestimated. Might it not be true that they still had no idea of these possibilities? Gamow then related the following story from his Soviet past. In 1932, before his eventual flight from the Soviet Union, he had referred at a scientific gathering to the work of Atkinson and Houtermans, in which, as was known, the fusion of light nuclei in the sun had first been suspected. After Gamow's lecture he had been approached by Bukharin, the People's Commissar, who had asked with interest whether he thought such reactions could not also be reproduced on the earth. Bukharin had even offered Gamow the use of all the current generated by the Leningrad electricity works for experimental purposes a few hours every night.*

Gamow's story strengthened the determination of Teller, Alvarez and Lawrence to force the construction of the Super bomb on the government as soon as possible. They promised one another to do everything in their power to attain this goal.

* According to a report by the German atomic physicists Gerlach and Joos, an atomic scientist had been found during the German invasion of the Soviet Union who was hoping to bring about a fusion of light nuclei by means of shaped charges. Unfortunately Gerlach and Joos do not remember the name of the Russian who had been carrying out these experiments.

Seventeen

Dilemma of the Conscience (1950-1951)

Hans Bethe had been famous among his colleagues and friends, throughout his life, for his invariable good humor and his—if possible—even better appetite. He was a healthy and happy man, inwardly and outwardly sure of himself, and could not by any stretch of the imagination be identified with the popular conception of an atomic scientist—a prey to conscientious hesitations. Yet this was the man who took it particularly hard when he, like other physicists, was faced with the question whether the hydrogen bomb ought to be built.

"I am unhappy to admit that—during the war at least—I did not pay much attention to this. We had a job to do and a very hard one," said Bethe, when he was later asked whether any moral scruples about the construction of atom bombs had occurred to him in Los Alamos. After Hiroshima his attitude changed. Like many other atomic scientists, he was troubled by the responsibility he bore for his share in the construction of this terrible weapon. As a member of the Emergency Committee of Atomic Scientists he took a leading part among those who urged that the public be enlightened on the danger of atomic warfare and insisted on the necessity of international control. He soon saw, earlier than most of his colleagues, that if the scientists wished to maintain their influence they would have to keep a certain distance from the turmoil of contemporary politics.

Bethe, as the son of a distinguished German physiologist, had been particularly reluctant to leave his native land in 1933. From Baden-Baden, where he was allowed to spend a few last enjoyable days before his final emigration, he wrote a most melancholy letter of farewell to his teacher, Sommerfeld, who had regarded him as his inevitable successor. From the moment of his arrival in the United States Bethe embarked upon a brilliant career, but he often looked back with longing to the old days when he had been obliged to live chiefly on a meager scholarship procured by his master. Sommerfeld asked him after the war whether he would care to accept the Chair of Theoretical Physics at Munich—it had been occupied after Sommerfeld's departure by the worst successor imaginable, a rabid adherent of German Physics named Müller. Bethe felt compelled to refuse. He had grown accustomed to his new homeland and above all considered himself under so deep an obligation to the Americans that he was no longer attracted by the highest goal of his former ambition, a professorship at one of the German universities.

His strong personality had created at Cornell University one of the most respected institutes of nuclear physics in the United States. That paradise of pure research was invaded, towards the middle of October 1949, by Teller, advocate of the "hell bomb." Teller intended to lead Dr. Bethe into temptation. He begged him to return to Los Alamos for just one year, since his collaboration in the production of the new weapon was indispensable.

Bethe was well aware of his own merit. He knew that Teller was not merely flattering him but really could not do without him. Bethe's brilliant colleague was rather like one of those Hungarian authors of boulevard plays whose excellent ideas are usually good enough for a magnificent first act but have seldom been thought out to the end. "Teller . . . needs . . . some control, some other person who is more able to find out what is the scientific fact about the matter. Some other person who weeds out the bad from the good ideas. . . ." Such was Bethe's judgment of his visitor, and he could see himself perfectly well in the part of the "other person."

When Teller found that Bethe was not to be tempted by financial offers, he tried to dazzle him with some glittering notions about the probable character of thermonuclear reactions. Bethe was, as he says himself, "very impressed by his ideas." The attraction of working with Teller, Ulam, Gamow and possibly also Fermi, especially with the aid of the greatly improved electronic computing machines which were at that time strictly reserved for military purposes, must have been extraordinarily strong. Plenty of interesting new discoveries could undoubtedly be expected from such an exceptional team.

Yet Hans Bethe hesitated. It seemed to him, as he immediately told Teller, "a fearful undertaking to develop an even bigger bomb." He had always talked over all important questions in life with his young wife, a daughter of the well-known German scientist Ewald. They discussed Teller's request until late that night. Later Bethe recalled that he had been deeply troubled about what he should do: "It seemed to me that the development of thermonuclear weapons would not solve any of the difficulties that we found ourselves in and yet I was not quite sure whether I should refuse."

Bethe acted as scientists do when they cannot solve a problem. He tried to discover more facts, especially political and military facts, before he might reach a final conclusion. He thought he would be best able to obtain them from Oppenheimer, who was undoubtedly in a position to form a more accurate estimate of the global situation than Bethe himself, since he served on a number of secret government committees.

Bethe had no sooner told Teller, who had stayed in Ithaca for the night, what he had decided to do, when the telephone rang. Oppie was on the line. He had heard of the efforts by Alvarez, Lawrence and Teller to convince the authorities of the necessity of a Super and wanted to know what Bethe thought. He had heard also that Teller was in Ithaca to discuss the matter. Oppenheimer proposed that the three of them should talk it over at Princeton.

This was precisely what Bethe wanted, but Teller suspected that Oppenheimer meant to argue either against him personally or against the Super. He appeared very depressed after this long-distance conversation. "I had been under the impression," he recalled a few years later, "that Oppenheimer was opposed to the thermonuclear bomb or to a development of the thermonuclear bomb. . . . I am pretty sure that I expressed to Bethe this worry, telling him, 'We are going to talk with Oppenheimer and then you will not come!' "

Two days later Bethe and Teller were sitting in the Director's office at the Institute for Advanced Study in Princeton. Oppenheimer had been in charge of the Institute since 1947 and regarded it as the most important of his many new preoccupations. There was a very great difference between the bright and orderly room in which they sat, with its outlook over broad meadows fringed with trees in all the varied tints of autumn, and Oppie's barrack-like office at Los Alamos, where the same three men had met during the war. In those days Oppenheimer had often seemed to his associates to resemble the enthusiastic founder and leader of some pioneer settlement in the Wild West. Now he reminded them of an English country gentleman receiving his guests at a stately home furnished with exquisite taste. In the same building, only a few doors away from Oppenheimer's office, Einstein, now in his seventies, was working in bare, unadorned surroundings at the splendid structure of the unified field theory embracing all the phenomena of gravitation, light and matter. He only rarely discussed scientific questions with the Director of his Institute. But whenever he read any news in his morning paper which did not please him, he used to call Oppenheimer on the house telephone with the indignant demand: "Now, what do you think of that?"

At bottom it was a simple question to which Bethe wanted an answer when he came to Princeton. But it was not answered for him. Oppenheimer showed him and Teller a letter which he had just received from James Conant. "Uncle Jim," as he was called in research circles, took a strong position against the new bomb project. He declared in that letter that if people absolutely must have it they could only get it over his dead body.

Apparently Oppenheimer did not share Conant's views, but he said nothing definite against them. He observed that if a hydrogen bomb were to be developed in the United States the work ought to proceed, from the start, with less secrecy than in the case of the atom bomb. He compared the United States to transparent glass and the Soviet Union to the semiopaque onyx. During the entire interview he refrained from any frank expression of opinion, either from caution in Teller's presence, or because he did not wish to say anything that might influence Bethe, or perhaps simply because he himself had not yet made up his mind.

Bethe was very disappointed with the turn the conversation had taken, so much so that after they had left Oppenheimer he said to Teller: "You see you can be quite satisfied. I am still coming."

But no sooner had Teller departed than Bethe's conscience again began to torture him. He consulted Victor Weisskopf, a close friend and colleague, who had been called the "Oracle" at Los Alamos. Since the end of the war Weisskopf had firmly declined to have anything more to do with atomic armaments. He was now teaching at M.I.T. and was considered one of the leading nuclear experts of his generation.

On a fine, warm autumn evening the two friends paced to and fro, deep in talk, until long after dark. Trees towered over and around them; a light wind was singing the fiery-red autumn leaves to sleep; a brook murmured a melody. Had mankind the right to destroy this kind of world, or even to imperil it? In 1939 Weisskopf had been one of the group with Szilard which had urgently demanded action, but he had learned from experience that when one gave soldiers a weapon they could hardly resist the temptation to pull the trigger.

The two men continued their conversation next day, during a drive to New York with Georg Placzek, a friend of both. He was not only an outstanding physicist but also a first-rate historian, with an especially deep knowledge of the Middle Ages. As they drove to New York through the monotonous industrial landscape, void of all tradition, the three European-born friends agreed that, as Bethe reported later, "after such a war, even if we were to win it, the world would not be such, would not be like the world we want to preserve. We would lose the things we were fighting for. This was a very long conversation and a very difficult one. . . ."

Bethe's struggles with his conscience were at an end. He was anxious to return to his university at Ithaca the same evening, but because of these highly important discussions he missed his plane. "Perhaps it's better so," he thought. "I ought to have another talk with Teller today."

It was difficult to find his colleague in the large city, but at last he reached him by telephone. Teller was at the home of Lewis Strauss, the only one of the five directors of the Atomic Energy Commission who approved, like Teller himself, a "crash program" for production of the Super. "Edward," said Bethe, "I've been thinking it over. I can't come after all."

On the morning of October 29, 1949, the Washington newspapers published some encouraging statistics. "Mortality in this city is at present lower than it has ever been," they announced. "It has fallen by about 25 per cent in the last ten years. This means that 15,000 of our fellow citizens and neighbors would not be alive today if medicine and hygiene had not made such encouraging progress."

The papers could not report that on the same day a debate was being held on the second floor of the Atomic Energy Commission building on Constitution Avenue to consider the question of constructing a weapon capable of increasing the mortality figure, almost in an instant, to between 80 and 90 per cent of the population affected. Only about a hundred people in the United States had any idea that on this day the General Advisory Commission, a body of nine leading American scientists, had met to come to a decision on the Super problem.

Since the beginning of 1947 the Commission had been sitting every few months, under the chairmanship of Robert Oppenheimer, who had been in charge of it from the start. On this occasion it had been summoned to answer a question put by the "supermen"—a term derived from the well-known comic strip—Lawrence, Alvarez, Teller and Strauss. They had asked: "Should the United States embark upon the production, as a matter of urgency, of a thermonuclear bomb?"

Oppenheimer opened the meeting by stating once more the matter to be debated. Then he asked each of the seven members present—one of the nine, Glen Seaborg, was abroad—to give their opinions in turn.* After they had all done so he stated his own view. None of the members spoke for longer than five or ten minutes. During the following days two reports were drawn up and discussed. They agreed on the point that the Super would probably be technically feasible, but that its production would be so extraordinarily complicated and uneconomic as to affect adversely the development program for fission bombs, which were being manufactured in the greatest variety of types and in growing numbers. From a military point of view it appeared doubtful whether there would be much point in constructing a Super, since there would only be two targets in the Soviet Union—Moscow and Leningrad—big enough to justify such a bomb. But in the third place—and this was the aspect on which by far the greatest emphasis was laid—all the members believed that the moral standing of the United States in the world would suffer if it developed such a weapon.

This view was expressed with special clarity and force by Rabi and Fermi in their joint memorandum. It stated:

> The fact that no limits exist to the destructiveness of this weapon makes its very existence and the knowledge of its construction a danger to humanity as a whole. It is necessarily an evil thing considered in any light. For these reasons we believe it important for the President of the United States to tell the American public and the world that we think it wrong on fundamental ethical principles to initiate the development of such a weapon.

* James B. Conant, President of Harvard, Lee Du Bridge, President of the California Institute of Technology, Enrico Fermi of the University of Chicago, I. I. Rabi of Columbia University, Hartley Rowe, President of the United Fruit Company, Oliver Buckley, President of American Telephone and Telegraph Company, and Cyril S. Smith of the University of Chicago.

Rabi and Fermi associated their rejection of the proposal to make the bomb with a suggestion that the President might make political use of a public repudiation of it by calling upon the Russians to agree to repudiate it in their turn. Any future breach of such agreement on the subject of thermonuclear weapons was to be regarded as justifying war.

The other six members of the Commission came to a more cautious but equally adverse conclusion:

We all hope that by one means or another the development of these weapons can be avoided. We are all reluctant to see the United States take the initiative in precipitating this development. We are all agreed that it would be wrong at the present moment to commit ourselves to an all-out effort towards its development.

In determining not to proceed to develop the super bomb we see a unique opportunity of providing by example some limitations on the totality of war and thus of eliminating the fear and arousing the hope of mankind.*

This victory of reason and moderation was permitted to live exactly three months. The "activists" stubbornly continued their campaign. They worked with success on the Air Force and the Chairman of the Joint Congressional Committee on Atomic

* Hans Bethe did not participate in the sessions of the General Advisory Commission, as he did not belong to it. It may be of some additional interest to note here the very decided views he expressed on the subject at the time. "I thought," he stated, "that the alternative might be or should be to try once more for an agreement with the Russians, to try once more to shake them out of their indifference or hostility by something that was promising to be still bigger than anything previously known, and to try once more to get an agreement at that time that neither country would develop this weapon. This is enough of an undertaking to develop the thermonuclear weapon that if both countries had agreed not to do so it would be very unlikely that the world would have such a weapon.

"Maybe the suggestion to negotiate again was one of desperation. But for one thing the difference was that it would be a negotiation about something that did not yet exist and that one might find it easier to renounce making and using something that did not yet exist than to renounce something that was actually already in the world. For this reason I thought that maybe there was again some hope. It also seemed to me that it was so evident that a war fought with hydrogen bombs would be destruction of both sides that maybe even the Russians might come to reason."

Energy, Brien McMahon. They broke through the defenses of Secretary Johnson, and Paul Nitze, head of the Planning Division of the State Department, who thought that it was absolutely necessary for the world to go on believing in the superiority of American technology. He advanced the opinion that such a belief alone was well worth the five hundred million dollars such a weapon was estimated to cost.

At last the advocates of the Super won over even Omar Bradley, Chairman of the Joint Chiefs of Staff, well known for his levelheadedness and moderation. His letter of January 13, 1950, in which he stated that he could not bear to think that the Russians might be the first to produce the hydrogen bomb and thus obtain a lead in the armaments race contributed more than anything else to bring about the already imminent change of feeling. It only needed one more shock to insure the support of the White House for the production of the Super.

That shock duly happened. On January 27, 1950, Klaus Fuchs left the English atomic research station at Harwell for London. He was met at Paddington Station by James William Skardon, a police inspector. The two men greeted each other amicably and then drove together to the War Office, where they entered one of the rooms and sat down. Skardon asked: "Are you ready to make a statement?" Fuchs nodded. He had known for some time that he was under suspicion. He now intended to make a full confession. He began: "I am Deputy Chief Scientific Officer (acting rank) at the atomic energy research establishment, Harwell. I was born at Russelsheim on the 29th December 1911. My father was a parson and I had a very happy childhood. . . ."

That same day the authorities in Washington learned that Fuchs had for many years been communicating to the Russians all the atomic secrets to which he had access. How much did he know? The Atomic Energy Commission, in response to an inquiry, was able to state on the following day that Fuchs had not only been supplied with information relating to the new, improved uranium bombs, but had also attended lectures and debates on the Super.

Fuchs told the inspector all about his activities as an agent. He refrained from giving him details of the technical information he had passed on, since Skardon had no right to knowledge of atomic data. He did not discuss that subject until January 30, when he talked about it exhaustively with Michael Perrin, the scientist appointed for that purpose, the man who had acted as wartime liaison officer for atomic affairs between the United States and Britain.

This sensational news could not, of course, fail to have its effect upon the General Advisory Commission, then again in session at Washington. On the following day, January 31, the Special Committee of the National Security Council appointed to deal with the Super problem met in the old building of the State Department, next door to the White House. The Committee consisted of Secretary Johnson, Secretary of State Acheson, Lilienthal, Chairman of the Atomic Energy Commission, and their associates. Deeply impressed by the Fuchs case, they resolved by two votes (Johnson and Acheson) to one (Lilienthal) to recommend that the President order a crash program to build the hydrogen bomb.

That same afternoon the American people, who had not been consulted in the matter, were informed of one of the greatest decisions in its history. President Truman solemnly declared: "I have directed the Atomic Energy Commission to continue its work on all forms of atomic weapons, including the 'hydrogen' or super bomb. Like all other work in the field of atomic weapons, it is being and will be carried forward on a basis consistent with the over-all objectives of our program for peace and security."

One of the hundreds of thousands of persons who read this alarming statement in their newspapers was Klaus Fuchs. At that moment he was still at liberty. On February 2, 1950, he agreed, after a telegraphic invitation from Perrin, to visit him at his London office in Shell-Mex House. Fuchs still believed that after making such a frank confession he would not be punished. He called at Perrin's office, as agreed, at precisely three in the afternoon. The police officer who had been ordered to arrest him there had not yet arrived because of a dispute about the wording of the warrant. He was a good fifty minutes late. Shortly after, Klaus Fuchs was on his way to Bow Street Police Station, the first of his prison residences.

The history of the relations and negotiations between the United States and Britain in the field of atomic affairs was, for the most part, still secret. The few who knew something about it noticed that the Fuchs case had come to light at the very moment that a British delegation in the United States was attempting to extend the scope, restricted for years, of the exchange of atomic information between the two countries. The arrest of Fuchs, who had been at Los Alamos as a member of a British delegation, at once led to the abrupt termination of the discussions in question, though they had had every prospect of success. The Americans now believed that British security measures for the protection of atomic secrets were too lax. Might it not be that the Russians had intended to bring about this state of affairs, and succeeded in doing so, by denouncing Fuchs themselves to the British Intelligence Service? It was some time since he had last given them any information. Had the Russians found a use for a man who had otherwise become of no value to them, as a weapon against closer Anglo-American co-operation?* If this is what the Russians intended, they certainly got what they wanted, but if so, they themselves had supplied the final impetus for the construction of the American "hell bomb."

This time public opinion was at last startled out of its mood of resignation. The "H-bomb," as it was thereafter called, aroused the same fear and indignation as the first atom bomb. Churchmen, scholars, politicians and editors throughout the world warned of the danger and called urgently for a new attempt to

* This supposition is strengthened by the fact that in the case of the Italian-born nuclear physicist Bruno Pontecorvo, who afterwards worked in England, the Russians chose to reveal his presence in Moscow, which had been concealed for years, at the very time when another British delegation was about to open negotiations in Washington for the release of atomic secrets.

reach an understanding between West and East. The American journalists Joseph and Stewart Alsop wrote: "The exploitation of the deepest secrets of creation for the purposes of destruction is a shocking act." Nobel prize winner Compton declared: "This is not a question for experts, either militarists or scientists. All they can do is to explain what the results will be if we do or do not try to develop such destructive weapons. The American people must themselves say whether they want to defend themselves with such weapons." Szilard stated in a broadcast that the radioactive effects of the Super bomb could be so much intensified that even the explosion of five hundred tons of heavy hydrogen would suffice to extinguish all life on earth. Einstein said with horror:

> The armament race between the U.S.A. and the U.S.S.R., originally supposed to be a preventive measure, assumes an hysterical character. On both sides the means to mass destruction are perfected with feverish haste—behind the respective walls of secrecy.
>
> If successful, radioactive poisoning of the atmosphere and hence annihilation of any life on earth has been brought within the range of technical possibilities. The ghostlike character of this development lies in its apparently compulsory trend. Every step appears as the unavoidable consequence of the preceding one. In the end there beckons more and more clearly general annihilation.

The leading spirit in the campaign against the hydrogen bomb was Bethe. He gave expression to one fear in particular. "It would hardly be possible today to eliminate the atom bomb from our armament program, for most of our strategy is based upon it. I should not care for the same situation to arise in connection with the H-bomb." An explanatory article written by Bethe for the respected periodical *The Scientific American,* dealing with the scientific, political and moral aspects of the Super bomb, contained the passage:

> I believe the most important question is the moral one: can we, who have always insisted on morality and human decency between nations as well as inside our own country, introduce this weapon of total annihilation into the world? The usual argument, heard in the frantic week before the President's decision and frequently since,

is that we are fighting against a country which denies all the human values we cherish and that any weapon, however terrible, must be used to prevent that country and its creed from dominating the world. It is argued that it would be better for us to lose our lives than our liberty; and this I personally agree with. But I believe that this is not the question; I believe that we would lose far more than our lives in a war fought with hydrogen bombs, that we would in fact lose all our liberties and human values at the same time, and so thoroughly that we would not recover them for an unforeseeably long time.

We believe in peace based on mutual trust. Shall we achieve it by using hydrogen bombs? Shall we convince the Russians of the value of the individual by killing millions of them? If we fight a war and win it with H-bombs, what history will remember is *not* the ideals we were fighting for but the method we used to accomplish them. These methods will be compared to the warfare of Genghis Khan, who ruthlessly killed every last inhabitant of Persia.

Several thousand copies of the issue in which this article appeared were confiscated and pulped by government agents, in defiance of the freedom of the press, on the pretext that the article revealed secrets of importance to national defense.

Bethe was also one of the twelve American physicists* who challenged President Truman's decision in a statement dated February 4, 1950:

We believe that no nation has the right to use such a bomb, no matter how righteous its cause. This bomb is no longer a weapon of war but a means of extermination of whole populations. Its use would be a betrayal of all standards of morality and of Christian civilization itself . . . to create such an ever-present peril for all the nations of the world is against the vital interests of both Russia and the United States . . . we urge that the United States, through its elected government, make a solemn declaration that we shall never use this bomb first. The circumstance which might force us to use it would be if we or our allies were attacked by *this* bomb. There can be only one justification for our development of the hydrogen bomb and that is to prevent its use.

* S. K. Allison, K. T. Bainbridge, H. S. Bethe, R. B. Brode, C. C. Lauritsen, F. W. Loomis, G. B. Pegram, B. Rossi, F. Seitz, M. A. Tuve, V. F. Weisskopf and M. G. White.

The American government gave no such reassuring promise, either then or at any later time.

The debate on the Super bomb renewed most acutely for many scientists the problem of their personal responsibility for the results of their work. This problem had been stated for the first time in most explicit fashion by the celebrated mathematician Norbert Wiener. He had been asked not long after the end of the war on behalf of the research department of an aircraft-building firm, which also produced long-range guided missiles, whether he would let the firm have a copy of a report he had written during the war at the request of a certain military authority. Wiener's reply included the passage:

> The experience of the scientists who have worked on the atomic bomb has indicated that in any investigation of this kind the scientist ends by putting unlimited powers in the hands of the people whom he is least inclined to trust with their use. It is perfectly clear also that to disseminate information about a weapon in the present state of our civilization is to make it practically certain that that weapon will be used.
>
> If therefore I do not desire to participate in the bombing or poisoning of defenseless peoples—and I most certainly do not—I must take a serious responsibility as to those to whom I disclose my scientific ideas.
>
> I do not expect to publish any future work of mine which may do damage in the hands of irresponsible militarists.

Wiener's radical attitude was decisively repudiated by most American scientists. They relied mainly on the counterargument of Louis N. Ridenour, in an answer to Wiener: "No one can tell what the result of any given scientific investigation may be. And it is absolutely certain that no one can prophesy the nature of any practical final product that may arise in consequence of such research. . . ."

To this constantly repeated objection the English crystallographer Kathleen Lonsdale has replied: "The risk that one's work, though good in itself, may be misused must always be taken. But responsibility cannot be shirked if the known purpose is criminal or evil, however ordinary the work itself may be."

Only a few scientific investigators in the Western world have in fact acted on this principle. Their honesty obliged them to risk their professional future and face economic sacrifices with resolution. In some cases they actually renounced the career they had planned, as did one of Max Born's young English assistants, Helen Smith. As soon as she heard of the atom bomb and its application, she decided to give up physics for jurisprudence.

A number of American scientific investigators hostile to armament work were to be found in the Society for Social Responsibility in Science. Its members differed from those belonging to other organizations in one decisive point: they were unwilling to wait until the politicians finally decided upon collective disarmament. On the contrary they expected every individual to take an immediate personal stand against the continuance of the atomic-armaments race.

One of the founders of this society, Professor Victor Paschkis of Columbia University, gives the following account of its history:

> In August 1947 I published in the *Friends' Intelligencer* [a Quaker periodical] an article entitled "Double Standards," in which I expressed my views of something I considered utterly unreasonable. This was the fact that scientists who were trying to amass funds for the enlightenment of the public on the dangers of atomic weapons were simultaneously continuing their work on the weapons in question. A. J. Muste, President of the Reconciliation Brotherhood, rang me up and said: "There must be other research workers who feel the same." . . .

This society expressed its horror at the development of arms technique in deeds as well as in manifestoes. It probably gained some members when it was announced that the United States intended to construct a Super bomb, but it never comprised more than about three hundred scientists in America though by 1950 Einstein and Max Born were members. Unfortunately they could exert little influence. They were even refused admittance to the organization comprising all the scientific bodies in America, the American Association for the Advancement of Science.* Protests soon died away. After a while no more was heard in public about the hydrogen bomb. Once again, "flaming indignation" had proved to be only a fire of straw.

In June 1950 the Korean War broke out. At once quite a number of scientists, hitherto reserved about co-operating in armament laboratories, returned to war research. They now considered it their patriotic duty.

One of them was no other than Hans Bethe. He hoped, as he said later, to convince himself by his work that the hydrogen bomb could not, in principle, be produced. Such an assurance would have seemed to him the best solution for the United States, which had far more to fear from a Super-bomb war than the Russians. Bethe finally played a decisive part in the ultimate production of the bomb he himself so feared and hated because of his outstanding erudition and systematic work. And—as the supreme irony—he was in the end actually entrusted with the task of writing its technical history.

In 1954, however, he said: "I am afraid my inner troubles stayed with me and are still with me and I have not resolved this problem. I still have the feeling that I have done the wrong thing. But I have done it."

Nature itself seemed, at the beginning of 1950, to be putting up a more successful resistance to Teller's plans than the atomic scientists who first protested against and then took part in the Super project. Immediately after the White House directive the Theoretical Division at Los Alamos had started calculations for the new bomb. Two groups tackled the problem independently. One made use of the first of the big electronic computing machines, the ENIAC, which had been constructed from von Neumann's plans, sent from Philadelphia to the artillery range

* The Society for Social Responsibility in Science had its own placement service at the disposal of scientists who lost their appointments because of refusing to undertake armaments work. Some found positions in underdeveloped countries, where they could apply their scientific knowledge to the struggle against famine and poverty.

at Aberdeen mainly to calculate ballistic curves. The second group was composed of only two men, Ulam and his assistant Everett. Their sole mechanical appliance was the ordinary calculator, which had also been used in calculations for the construction of the first atomic bombs.

This system of working out the same problem in two groups, which could then compare the results independently arrived at, was already traditional at Los Alamos. It was practiced there, candidly, as a kind of intellectual sport. Rolf Landshoff, an emigrant from Berlin to the United States, who had belonged to Teller's group during the war, remembers in connection with this "racing" that "there was a meeting in Teller's office with Fermi, von Neumann and Feynman in which I took part because I was to carry out the calculations planned at that meeting. Many ideas were thrown back and forth and every few minutes Fermi or Teller would devise a quick numerical check and then they would spring into action, Feynman on the desk calculator, Fermi with the little slide rule he always had with him and von Neumann in his head. The head was usually first, and it is remarkable how close the three answers always checked."

In the case of the calculations for the Super the handicap that Ulam had assumed seemed almost too heavy for him. It was supposed that he would not be ready for some days or even weeks after the ENIAC. But, as is well known, these artificial brains talk a language of their own, into which any problem put to them must first be translated. Such programing is seldom free from error. The machine "notices" the fact and gives senseless answers which indicate, after close study, where the error lies.

This process all took time, which Ulam well understood how to employ. Before the ENIAC group had come to the end of the fault-detecting period and put their corrected questions to the electronic oracle, Ulam, by taking a few bold short cuts, had already reached the goal and submitted his results. They were discovered to be, if correct, fatal to Teller's plans. According to these data the hydrogen bomb, as hitherto conceived, would either be utterly impracticable or could only be produced with so great a quantity of the rare hydrogen isotope tritium that its cost would apparently be far too high.

Teller reacted to this news like an oriental despot. He couldn't very well behead Ulam, that bringer of evil tidings, but he caused him to fall from favor. When, soon afterwards, the first results came in from the ENIAC group and appeared promising, the suspicious Teller imagined Ulam might have deliberately deceived him. After all, there were a number of people in Los Alamos who were only working there because they hoped that the Super would turn out to be impossible. But a little later, further results from the big Aberdeen computer brilliantly confirmed those of the Polish mathematician. They ratified his calculations in every detail.

There it was in black and white, stated as a mathematical certainty. All the work hitherto done on the Super had been, in Teller's own words, "nothing but fantasies." It had to be started all over again. Had the preliminary measurements themselves, upon which the calculations had so far been based, in fact been accurate? One could find out only by testing them afresh in actual trial. If practical results were to be obtained, much more precise observations would have to be taken in the new test than in any previous undertaking in the atomic-armaments field. Instruments of hitherto unknown speed and precision were essential. Cameras would have to take thousands of photographs in the fraction of a minute. A system of signals would be necessary to relay their "experiences" to a distant control point before they themselves were destroyed by the force of the explosion. Countless artificial organs, electronic eyes, ears and noses, superior to the corresponding human senses, would have to deliver data to a laboratory set up on the remote atoll of Eniwetok in the South Pacific. Such data might then suggest to the theoreticians a new method of procedure which would have a chance of success.

The test for which Teller and his aides were preparing between 1950 and the middle of May 1951 bore the code-name "Greenhouse." They themselves called it far more often—and in certain respects more appropriately—"Icebox." The monstrous device which they intended to send up into the air had to be kept at a very low temperature to enable the heavy hydrogen, or tritium, to maintain the state of aggregation required for an explosion of such magnitude. Much later this most expensive and grandiose of all Super-bomb tests was given the nickname "Superfluous." Although a rich haul of experimental results was obtained, they were eventually found to have little bearing on the solution of the Super crisis it had been hoped they would provide.

Before the experiment was carried out Stan Ulam, whose calculations had reduced to absurdity the original plans for the hydrogen bomb, picked up an altogether new scent. He communicated his idea, which pointed in a wholly different direction, to Teller, who had meanwhile apologized for his earlier suspicions. Though Teller was at first unwilling to follow this line of research, he eventually adopted it. He began by discussing the suggestion with Frédéric de Hoffman, his young assistant. Hoffman recalls that he thought nothing of it at the time "because, after all, Edward is always having an idea. But the next morning he came in to me and said, 'Freddie, I think I really have something. Stick some figures into it.' He told me about it and I started to work with my desk calculator. The answer came out right."

This suggestion originally made by Ulam was responsible for the development of that ingenious idea which finally made possible the construction of the American Super. It was in June 1951 that Teller revealed his idea for the first time to a larger number of experts, assembled at the Institute for Advanced Study for a weekend debate on the existing state of the "thermonuclear question."

The intellectual climate had greatly changed since the October days in 1949, when the majority of those now gathered had declared themselves, mainly on political and ethical grounds, opposed to the construction of Super bombs. The change is evident in the report of an eyewitness, Gordon Dean, then Chairman of the Atomic Energy Commission:

We had at that meeting in June of 1951 every person, I think, that could conceivably have made a contribution. People like Norris Bradbury, head of the Los Alamos laboratory, and one or two of his assistants, Dr. Nordheim, I believe, was there from Los Alamos, very active in the H program. Johnny von Neumann from Princeton, one of the best weapons men in the world, Dr. Teller, Dr. Bethe, Dr. Fermi, Johnny Wheeler, all the top men from every laboratory, sat around this table and we went at it for two days.

Out of the meeting came something which Edward Teller brought into the meeting with his own head, which was an entirely new way of approaching a thermonuclear weapon.

I would like to be able to describe that but it is one of the most sensitive things we have left in the atomic-energy program . . . it was just a theory at this point. Pictures were drawn on the board. Calculations were made, Dr. Bethe, Dr. Teller, Dr. Fermi participating the most in this. Oppy very actively as well.

At the end of those two days we were all convinced, everyone in the room, that at least we had something for the first time that looked feasible in the way of an idea.

I remember leaving that meeting impressed with this fact, that everyone around that table without exception, and this included Dr. Oppenheimer, was enthusiastic now that you had something foreseeable. I remember going out and in four days making a commitment for a new plant . . . we had no money in the budget to do it with and getting this thing started on the tracks, there was enthusiasm right through the program for the first time. The bickering was gone. The discussions were pretty well ended, and we were able within a matter of just about one year to have that gadget ready.

This report does not sound as if it were concerned with men who had abandoned only their "buts" with reluctance, after long inward conflict. How is one to explain such macabre enthusiasm, which had swept away all the earlier scruples and objections to the Super monster? Oppenheimer himself provides a clue to the reason why scientists of today, despite occasional hesitations, in the end so often change their minds when the successful solution of a problem they have long wrestled with is at last in view, however disastrous its ultimate effects may be. In recalling the repudiation of the hydrogen bomb by the General Advisory Committee in October 1949 he said:

> I do not think we want to argue technical questions here and I do not think it is very meaningful for me to speculate as to how we would have responded had the technical picture at that time been more as it was later.
>
> However, it is my judgment in these things that when you see something that is technically sweet you go ahead and do it and you argue about what to do about it only after you have had your technical success. That is the way it was with the atomic bomb. I do not think anybody opposed making it; there were some debates about what to do with it after it was made. I cannot very well imagine if we had known in late 1949 what we got to know by early 1951 that the tone of our report would have been the same.

In this statement there is no longer any trace of the ethical doubts so forcibly expressed in the report of the General Advisory Committee. Oppenheimer here, whether intentionally or not, reveals a dangerous tendency in the modern research scientist. His remarkable admission perhaps explains why the twentieth-century Faust allows himself, in his obsession with success and despite occasional twinges of conscience, to be persuaded into signing the pact with the Devil that confronts him: What is "technically sweet" he finds nothing less than irresistible.

Eighteen

In the Sign of the "MANIAC" (1951-1955)

After that memorable weekend at Princeton, the direction of the road which would presumably lead to the Super was known, but it was blocked at the start by still another almost insurmountably high mountain range of figures. Even the calculations relating to the atom bomb had involved thousands upon thousands of detailed computations. The precise determination of a thermonuclear explosion would be many degrees harder, for a physical process comprising numerous stages would have to take place in a fraction of a second. The steps must be foreseen with the greatest possible accuracy. On the basis of these assumptions an infinitely complex apparatus would have to be constructed.

All this work had to be done at an even higher speed than had been necessary in the Second World War. Eighteen months had already gone by since President Truman's directive for the construction of the Super. The Russians were probably devoting all their energies to the same problems.

Teller and the Director of the Los Alamos laboratory, Norris Bradbury, mobilized their whole forces for the conquest of the mathematical Mount Everest. Workers at the laboratory immediately resolved to work six instead of five days a week, while the computer section actually introduced day and night shifts.

Cerda Evans, a specialist in the field of the new "electronic brains," states:

> Never in my life have I been obliged to sleep and breakfast at such impossible hours as during those months, when we sat at our computers for twenty-four hours a day, relieving one another at intervals. The ENIAC on which we worked, though faster than any other previous mathematical apparatus, was temperamental and delicate. Some tube or other or some circuit was forever going wrong. On those occasions we simply had to wait. Once a storm put the mechanism right off balance. We all sat glued to the telephones in our rooms, waiting for the repair crew to report that we could carry on. Several times they called us up to say that we could come over, as everything would be all right in ten minutes. But when we rushed to the spot, it would turn out to be only another false alarm. So it went on for a whole week.

Until every one of the calculations had been completed there was simply no possibility of any real progress. But they took so long that one could see no end to them. Once more a crisis seemed to be at hand. At that moment the situation was saved by the mathematician and atomic scientist John von Neumann, who told Teller he hoped to have a new electronic computer ready within a few months which would be incomparably more effective than the ENIAC.

Even during his years as an undergraduate at Göttingen the ingenious Hungarian von Neumann had been nicknamed "Dr. Miracle" by his fellow students, because of his passion for mechanical toys. They were thinking of E. T. A. Hoffmann's weird builder of automata, who invented the life-size and life-like doll Olympia and then fell desperately in love with it.

In 1930 von Neumann, already considered one of the leading mathematicians of his generation, had emigrated to the United States. At first he felt far from comfortable in the New World. Von Neumann, apart from his mathematics, was fonder of a free and easy social life than of anything else. At Princeton, however, there were no cafés, like those in Central Europe, where one could gossip and argue for hours over a cup of coffee. The scholarly man missed this institution so much that he began to wonder in all seriousness whether he should invest his rather meager fortune in a business of the kind. "But, Johnny," his American colleagues objected, "the citizens of Princeton wouldn't know what to do with a Viennese café!" "Don't bother about that," von Neumann replied. "We'll recruit a few of our European colleagues. They'll sit in my café every afternoon for a few days, just to show you how it's done."

This plan, which was eventually given up, met with less approval among von Neumann's new countrymen than his passion for robots. The recent advances in electronics made in the United States favored the cultivation of his hobby. Soon he was devoting more and more time to it. The similarities between human beings and machines fascinated him. He proceeded to invent a whole series of mechanisms with human or even superhuman properties.*

Teller had experienced no difficulty in winning over his compatriot von Neumann, from the start, to take part in the Super project. Unlike Oppenheimer and other atomic scientists, who had at first suffered from conscientious scruples about the new bomb, the Hungarian mathematician had immediately declared his approval of it, because he feared nothing in the world so much as Communism. He had learned to hate it as a thirteen-year-old boy during its brief period of domination in Budapest after the First World War. There were few experiences in his life that had left such a deep impression on him as those days of terror and his flight from the city. Since then he had adopted, at any rate when Bolshevists were being discussed, a hard-boiled attitude.

Von Neumann immediately saw what an indispensable part his new computer would be able to play in the production of the "hell weapon." He did all he could to speed up its construction,

* In this connection von Neumann thought up, among other models, one capable, so long as it was supplied with enough raw material, of continuous self-reproduction. It was to consist of a box and a "genetic tail" which contained the basic elements of its posterity. Von Neumann's pupil Kemeny states: "One could further arrange to limit the supply of raw material, so that the machines would have to compete for 'Lebensraum' even to the extent of killing one another."

while at the same time his pupils Nicholas Metropolis and James H. Richardson built up an identical computer at Los Alamos.

The ENIAC, which could only remember twenty-seven "words," had a sparrow's memory in comparison with that of the new electronic brain. This could retain 40,000 bits of information at once and, if necessary, recall them later. It was so accurate that it could check the instructions given it, identify errors and, if given the chance, correct its faulty orders. When von Neumann released his last invention for use, it aroused the admiration of all who worked with it. Carson Mark, head of the Theoretical Division at Los Alamos, recollects that "a problem which would have otherwise kept three people busy for three months could be solved by the aid of this computer, worked by the same three people, in about ten hours. The physicist who had set the task, instead of having to wait for a quarter of a year before he could get on, received the data he required for his further work the same evening. A whole series of such three months' calculations, narrowed down to a single working day, were needed for the production of the hydrogen bomb."

It was a calculating machine, therefore, which was the real hero of the work on the construction of the bomb. It had a name of its own, like all the other electronic brains. Von Neumann had always been fond of puns and practical jokes. When he introduced his machine to the Atomic Energy Commission under the high-sounding name of "Mathematical Analyzer, Numerical Integrator and Computer," no one noticed anything odd about this designation except that it was rather too ceremonious for everyday use. It was not until the initial letters of the six words were run together that those who used the miraculous new machine realized that the abbreviation spelled "maniac."

Work with the MANIAC went on with a good deal less friction than between the Los Alamos team and another great brain, Edward Teller. Just as he had failed to adjust during the Second World War, he constantly tried to direct the tempo and method of Director Bradbury's own activities. Teller hinted to his influential friends in Washington that the leading personalities on the mesa still paid too much attention to Oppenheimer's views and were accordingly more concerned with the production of better atom bombs than with the construction of the hydrogen bomb. From Teller's dissatisfaction came the idea of building a second nuclear-weapons laboratory, in addition to Los Alamos, with himself as its boss. It was to be devoted exclusively to the problems of a thermonuclear bomb.

The idea found special favor with the Air Force. At that time, in 1952, this branch of the armed services feared that it might have to share with the other two branches, but in particular with the Army, its monopoly of employing the atom bombs. The General Advisory Commission, led by Oppenheimer, several times rejected the idea of the second laboratory as unnecessary, but in the summer of 1952 the Commission was outmaneuvered. Preparations were begun to expand a small laboratory which had been used only occasionally for research for the University of California.

The little town at which this new smithy for the forging of atomic weapons arose was called Livermore. By an irony of fate, it had been founded by Robert Livermore, a veteran weary of the naval battles against Napoleon. As an ordinary seaman he had deserted the British warship *Colonel Young* at the California port of Monterey. His wanderings led him in 1835 to a green valley which reminded him of the landscapes in central Italy. He married a local girl, had eight children and developed his property, Las Positas, into a flourishing estate. In 1952 bulldozers invaded this idyllic retreat in the Golden West, and a few months later the thermonuclear laboratory of the Atomic Energy Commission was built. Teller's departure from Los Alamos, which had for some time been inevitable, came in July 1952. Soon after, with E. O. Lawrence and Herbert York, he took charge of the new installations.

Meanwhile, on the Hill, the construction of the first Super bomb was approaching completion, though its spiritual father had gone. That autumn Marshall Holloway, director of this final phase of the work, arranged for some new apparatus at Los Alamos, in which a quarter of a billion dollars in taxes had been invested for expansion and technical equipment since 1945. The new machinery practically eliminated risk from the most dangerous part of the work, the determination of critical mass inside the bomb.

Experiments were now no longer carried out with such primitive resources as had been available in the time of Louis Slotin but with the help of a critical assembly under remote control, which bore the name "Jezebel." It had been placed, with two other similar devices, "Topsy" and "Godiva," behind heavy radiation shielding in two flat-roofed buildings so "hot"—*i.e.*, highly radioactive—that they could be entered only if special precautionary measures were taken. The control room, from which the machinery was guided, stood in the main laboratory, a quarter of a mile from the danger zone. What went on inside the "kivas" was observed only on television screens. The buildings had been called Kivas after the sacred ceremonial chambers of the Pueblo Indians, whose priests approached them with the greatest awe.

The thermonuclear device eventually produced by Teller and von Neumann, about a hundred devoted scientists, MANIAC and Jezebel, was not yet an actual projectile but a thermonuclear device, weighing no less than 65 tons. The tritium it contained, among other components, was an artificial hydrogen isotope produced in the uranium pile which had to be maintained at a fixed low temperature in a freezing apparatus as heavy as it was complex.

At the beginning of October 1952 thousands of scientists, test engineers, mechanics, soldiers and sailors gathered on what had become a nuclear testing ground, the atoll of Eniwetok, one of the Marshall Islands, a U.N. trusteeship territory under U.S. supervision. They prepared to shoot off "Mike," as the monster was called. Before the test Vannevar Bush, who had directed American research during the Second World War, made a strenuous attempt to induce the government to initiate negotiations with the Russians before taking this fresh step into a "hideous sort of world." His advice was declined. The device was installed on the islet of Elugelab. It was placed in a big protective shed whose massive rectangular shape reminded some of the participants of the Kaaba, the building which houses the sacred stone of the Moslems at Mecca.*

On the night of October 31-November 1, 1952, a final roll call of all the personnel was held. Roy Reider, the security director, had insisted as a precautionary measure that all the islands be evacuated. The people were taken aboard the waiting ships. In such experiments one must always count on, for safety's sake, a bang ten times as big as calculated. In this case one had to take into consideration also the possibility of an upheaval of the sea. All except a small group of specialists went at least forty miles away from the point of the expected explosion. The experts stayed on Elugelab to set off the bomb shortly before dawn. Reider recalls that these men were "as lonely as a group of wandering lepers, though at no time was this party ever out of communication with headquarters back at the control center." As soon as these specialists had also been taken to a place of safety the counting of minutes and seconds over the ships' loudspeakers began. Every eye stared into the distance out of which the light of the first man-made star would rise.

Edward Teller had received a formal invitation from Bradbury to attend the test of the Super in the Pacific. But Teller had, for understandable reasons, declined. About a quarter of an hour before the great event—it was then just before midday on the West Coast of the United States—he was walking slowly, with

* In connection with the choice of this locality for the first test of the hydrogen bomb the American author and painter Gilbert Wilson noted a strange coincidence. While he was reading *Moby Dick* it struck him that "only a century after Herman Melville wrote his great book our own American atomic engineers unwittingly selected almost the very spot in the broad Pacific, some few thousand miles south-east off the coast of Japan, where the fictional *Pequod*"—this was the American whaler commanded by the vengeful and fanatical Captain Ahab—"was rammed and sunk by the White Whale. . . . Melville had Ahab describe the Whale with an image remarkably similar to the conventional symbol of the atom used by artists, 'O trebly hooped and welded hip of power!' "

bent head, along a narrow path across a park on the grounds of the University of California at Berkeley, on his way to Haverland Hall, a building where one of the most sensitive seismographs in the world was kept in the basement. There Teller hoped to see signs of the shock initiated by the experiment five thousand miles away. The small room, with the sensitive instrument embedded in its floor of rock, was lit only by a single red lamp. It was turned off. Teller was left alone with a loudly ticking kitchen clock and the recording apparatus, which could indicate the slightest tremor on a photographic plate, with a beam of light a millimeter thick. He relates what happened then:

After my eyes became accustomed to the darkness I noticed that the spot seemed quite unsteady. Clearly this was more than what could be due to the continuous trembling of the earth, to the "microseisms" that are caused by the pounding of the ocean waves on the shores of the continent. It was due to the movements of my own eyes which in the darkness were not steadied by the surrounding picture of solid objects. Soon the luminous point gave me the feeling of being aboard a gently and irregularly moving vessel, so I braced a pencil on a piece of the apparatus and held it close to the luminous point. Now the point seemed steady and I felt as if I had come back to solid ground again. This was about the time of the actual shot. Nothing happened or could have happened. About a quarter of an hour was required for the shock to travel, deep under the Pacific basin, to the Californian coast. I waited with little patience, the seismograph making at each minute a clearly visible vibration which served as a time signal. At last the time signal came that had to be followed by the shock from the explosion and there it seemed to be: the luminous point appeared to dance wildly and irregularly. Was it only that the pencil which I held as a marker trembled in my hand? I waited for many more minutes to be sure that the record did not miss any of the shocks that might follow the first. Then finally the film was taken off and developed. By that time I had almost convinced myself that I must have been mistaken and that what I saw was the motion of my own hand rather than the signal from the first hydrogen bomb. Then the trace appeared on the photographic plate. It was clear and big and unmistakable. It had been made by the wave of compression that had traveled for thousands of miles and brought the positive assurance that Mike was a success.

A crater a mile long and 175 feet deep had been dug in the Pacific. As soon as the fireball of the first Super, a flaming dome three and a half miles in diameter, had disappeared and the vast, mushroom-shaped cloud of smoke rose into the sky, the observers realized a fact which they at first could hardly believe. The island of Elugelab had disappeared. The shot, in which the energy of three megatons—three million tons—of TNT had been liberated, had, like the first atom bomb, surpassed all expectations, even the calculations of MANIAC.

Mike's successful explosion had given practical proof, for the first time, that it was possible to reproduce on earth processes presumed to take place in the sun. The "monster" was not yet a bomb which could be carried in an aircraft. The American scientists in the armaments race were not really satisfied with even this notable technical success. They decided to start work, at the same speed, on a "dry" type of bomb, which would act on an isotope of lithium instead of tritium and thus dispense with the refrigerating apparatus. They were still engaged, at Los Alamos and Livermore, on the production of this new bomb, known simply by the affectionate name of "The Sausage," when news from Soviet Russia startled scientists all over the world.

Malenkov, Stalin's successor, proclaimed on August 8, 1953, that "the United States no longer has a monopoly of the production of the hydrogen bomb." Four days later the Radiation Detection Patrol identified traces, in the skies over Asia, of a new Soviet bomb explosion. Samples of these traces were subjected to laboratory analysis. The results caused an excitement among the initiated which could only be compared with that following the news of the first Soviet bomb. The radiation chemists reported that the Russians already possessed the "dry bomb." *

* The well-known Austrian physicist Hans Thirring had already foreseen the possibility of such a bomb in his book *The History of the Atom Bomb,* which appeared in 1946 in Vienna. In that work he wrote: "Nor

They were now probably in a position, as Sterling Cole, Chairman of the Congressional Atomic Committee explained with considerable alarm to a small group of his colleagues, to threaten the United States with hydrogen bombs at any time, while the Americans could for the present retaliate only with atom bombs. But the American government did not communicate these facts to the people.

The conditions which Washington had so long feared had now actually come to pass. The other side now held the lead in the race for the "absolute weapon." What could be done to overtake and if possible pass it? The race now accelerated to positively breakneck speed. To gain a few minutes' grace before the infliction of a hostile onslaught that might kill millions of people and cripple American industrial establishments at a single blow, a beginning was made on an "electronic wall" designed to approach the North Pole and reach far out to sea. To save a few minutes an idea was revived which had been given only casual attention since 1945—to put into operation unmanned missiles guided by remote control, capable of crossing the Atlantic or the snowy wastes of the North in less than a half hour.

These "intercontinental ballistic missiles" had been planned mainly under the direction of Germans, who had built the V2 weapons, and who now had become naturalized Americans. The main reason why these machines had not been built in great numbers was because it was calculated that their deviation from the target over a range of 5,000 miles would amount to 1 per cent, or fifty miles. Improvement of the control mechanism succeeded eventually in reducing the margin of error to .2 per cent. But even that meant a deviation of ten miles. Consequently, if such a missile were aimed at Moscow, it would not burst in the center of the city but in one of the suburbs. And if a long-

---

is lithium, in fact, by any means a rare element. Consequently, in a super atom bomb, about as many tons of lithium hydride could be used as there are kilograms of plutonium in the present bomb. Under such conditions an effect some thousands of times greater than that hitherto known could be achieved. God help any country over which a six-ton lithium hydride bomb explodes."

range rocket were aimed at a Leningrad airfield it would fall in open country or even in the sea.

Nevertheless, the Air Force believed that this sinister air torpedo, if it could only be more accurately aimed, would be the card with which the latest Russian ace could be most expeditiously trumped—unless it was even too late for that. Reports of the U.S. Central Intelligence Agency had already begun to inform the Pentagon of amazing Russian advances in precisely that same field of long-range guided missiles.

John von Neumann, then already suffering from an incurable cancer, was placed at the head of a secret committee to study this problem. The committee met for the first time in September 1953, a month after the announcement of the Russian H-bomb explosion. Von Neumann had in the past been closely associated with the Atomic Energy Commission and became a Commissioner shortly afterwards. He was therefore aware that plans for a "three-stage" bomb had been drawn up by the Commission. They wanted to add to the two stages of a thermonuclear bomb (that is, the central atomic-bomb explosion which initiates the fusion reaction in the thermonuclear fuel surrounding it, so producing a much greater effect) a third stage consisting of the fusion of the bomb's casing. This was to be made of U238 instead of ordinary material. This bomb, named the "Fission-fusion-fission" or FFF bomb, had not hitherto been constructed, the only reason being that it was considered likely to cause the maximum of ruin and destruction. Its radioactive fission matter would be spread over an area of 300 square miles. The weapon was therefore regarded as "overdestroying."

But von Neumann now proceeded to work out in his mind a combination of the intercontinental missile's uncertainty of aim with the enormous dispersion area over which the effect of the three-stage bomb would operate. He united two monsters, hitherto considered impracticable for war purposes, in unholy matrimony and presented them to his grateful employers as the "absolute weapon." For now, even if the intercontinental missile burst a long way from its target, it would still include it in its circle of death and annihilate it. The American periodical

Fortune reported later on von Neumann's ingenious compound: "Because of the quantum jump in the destructive power of the thermonuclear warhead, not to mention the still greater area of lethal fall-out, delivery within eight or ten miles of the center of the target became militarily acceptable."

Von Neumann, "the mathematician who gave the green light," as *Fortune* put it, is the clearest instance of a scientist who developed from an inventor of new weapons into a scientific strategist. During the years following the end of the Second World War, wrote Hanson W. Baldwin, an American expert on military subjects, "the technological revolution has made its major strides and has produced by far the greatest effects upon the art of war in the whole century and a half since Napoleon." The atomic scientists had begotten an entire family of nuclear weapons of all sorts and sizes. The specialists in aerodynamics and the aircraft engineers had built jet fighters capable of unprecedented maneuvers and unheard-of speeds. The rocket experts had developed tremendously fast missiles in great variety. In addition, as Baldwin explains, "biological and chemical weapons such as radioactive dusts and gases" had been made practicable.

Advances in electronics enabled all these messengers of death to be aimed accurately even at the highest speeds. But the generals, if left to their own resources, would never have been able to keep pace mentally with this "quantum jump" in the technique of war. They needed scientists at their elbows to help them work out their plans and revise them when it became necessary with every step forward in technology. Most of the scientific experts who took part in this war game continued to do so under the compulsion of the anxiety and suspicion to which they had been a prey ever since 1939. They participated in these progressively more and more desperate maneuvers of increasing threat and counterthreat, because of their belief that it was only by so doing that they could hope to preserve peace.*

* The American psychiatrist Lawrence S. Kubie, in an inquiry into the deeper motives of the scientists who took part in the armaments race,

The planning of future wars of this kind—in the hope that they would never take place if only they were prepared beforehand—also gave the MANIAC and other electronic "oracles" made in its image new parts to play as indispensable instruments for extremely rapid assessments and decisions in the strategic field. For the MANIAC even the end of the world was only one more question to be answered by calculation. If a city or a nation were translated into the equivalent expressions in the language of the machine, history and life would converge in a mass of dry figures and convenient formulae could be worked out for the annihilation of millions of people. When von Neumann's pupils or their opposite numbers in the Soviet Union sat at a computer panel working out the chances of war, they divided the resources of a country by the panic and despair of its population or multiplied them by its inventive capacity and obsession with victory. Would the problem be solved only if the loser were reduced to zero? Or should one simply proceed to retaliation by degrees instead of to radical retaliation? In such calculations it was always millions of deaths that were involved, from one million to hundreds of millions. And the time to elapse before annihilation became imminent grew ever shorter. If one could only forget—and one did gradually forget it almost completely, sitting at the console of such an instrument—that the single lives of human beings were at stake, the whole business could be regarded in principle as nothing more than one of those necessary calculations of probability for ascertaining the behavior of the millions of atomic particles in the inner mechanism of the new bombs.

Yet all these calculations that worked with such furious precision were suddenly interrupted by a perfectly ordinary sea breeze which no one had thought of considering. The weather forecast

---

advanced a different interpretation in the *The American Scientist* for January 1954: "Are we witnessing the development of a generation of hardened, cynical, amoral, embittered, disillusioned young scientists? If so, for the present the fashioning of implements of destruction offers a convenient outlet for their destructive feelings; but the fault will be ours and not theirs if this tendency should increase through the coming years and should find even more disastrous channels of expression."

on March 1, 1954, stated that it seemed to be moving north from the atoll of Bikini. But then it quite unexpectedly turned south, drifting over the islands of Rongelap, Rongerik and Uterik, till it reached the open sea where a Japanese steam trawler, the *Lucky Dragon No. 5,* lay. The boat was overtaken by a "snowstorm" out of a clear sky. It was not until two weeks later that the world learned the storm had been a rain of radioactive ashes. The tiny particles of dust found by Japanese scientists in the seams of the ship's planking contained the secret of the three-stage bomb, exploded for the first time on March 1, 1954, as "shoot one" of a new series of H-bomb tests.

At this very time the danger of war had once more become particularly acute. The French military post at Dien-Bien-Phu in Indochina was about to fall. In Washington and Paris the question of American intervention against the advancing Indochinese Communist Army was being considered. Admiral Radford, Chief of Staff, had just proposed that a "tactical atomic bomb" should be employed.

But it never came to that. Just as in August 1945 and February 1950, so now, for the third time, the entire world was seized with horror at the frightful violence of the new weapons. The Japanese fishermen had been far beyond the danger zone determined by the Americans. And yet they had been exposed, some 120 miles away from the point of explosion, to its effects. They reached their home port of Yaizu on March 14, sick and weak with sufferings they could not account for, and were at once taken to the hospital.

It was rumored that the scientists had lost control of the new bomb, which had liberated the terrific quantity of energy equal to between 18 and 22 million tons of TNT. Mike's explosive force had been equal only to 3 million tons of dynamite. It was admitted that the bang had been twice as powerful as had been anticipated. But even more disturbing than this news was the poisonous effect of the new projectile, which was identified during the following days in rain over Japan, in lubricating oil on Indian aircraft, in winds over Australia, in the sky over the United States and as far away as Europe.

The previous bombs had affected only the conscience of

mankind, so soon to relapse again into apathy; but the latest "hell bombs," it was evident from the reports, endangered the air that man breathed, the water he drank and the food that he ate. They menaced, even in times of peace, the health of every person, wherever he lived.

Admiral Strauss naturally lost no time in mounting a counteroffensive. He announced that his own scientists considered it an exaggeration to fear that the spread of increased radioactivity would endanger life wherever it was to be found. A dispute ensued between "appeasers" and "alarmists" which seemed likely to last for a good many years, for the most dangerous consequences of the radioactivity disseminated by the tests—its effects on posterity—could not be accurately assessed by science for generations.

All geneticists are agreed on the fact, though not on the extent, of the risk to the health of those who come after us owing to the bomb's liberation of cell-poisoning material. The criticism of Strauss expressed by the distinguished American specialist in heredity A. H. Sturtevant was particularly searching. He wrote:

> There is no possible escape from the conclusion that the bombs already exploded will ultimately result in the production of numerous defective individuals—if the human race itself survives for many generations. . . . I regret that an official in a position of such responsibilty should have stated that there is no biological hazard from low doses of high energy radiation.

Not long afterward the same scholar said in a public address that probably 1,800 of the children born in 1954, the year of the bomb test, were already infected by the high radiation. In the same year the American zoologist Curt Stern declared: "By now everyone in the world harbors in his body small amounts of radioactivity from past H-bomb tests: 'hot' strontium in bones and teeth, 'hot' iodine in thyroid glands."

The American physicist Ralph Lapp uttered an even more serious warning, based on conversations with an A.E.C. biologist, who would not allow publication of his name for fear of dismissal. Lapp stated:

> In 1945 55,000 tons of TNT equivalent were detonated. Sporadic testing followed until the tentacles of the cold war gripped the na-

tion and the test rate zoomed up. By 1954 the test rate was a thousand times greater than in 1945. My calculations show that if the world does follow the test schedule I have assumed by 1962 there will be an amount of radiostrontium committed to the stratosphere which some time in the 1970's will add up to the "maximum permissible amount" for the world's population.

Just what do we mean by the maximum permissible amount of a radioactive substance? Will more than this amount cause illness or death? If so, how much more? Is it safe for all humans to have one MPA (maximum permissible amount) of strontium retained in their bones? All MPA limits in the past have been specified for small groups of healthy adults working under controlled conditions and exposed to known hazards.

For such people the International Committee on Radiological Protection has laid down definite occupational MPA's. There is considerable disagreement among authorities on what MPA's should be for whole populations.

In the specific case of radiostrontium it is well known that growing children are more sensitive to strontium injury than adults. Therefore most experts believe that the MPA for whole populations should be ten times less than the occupational level.

Radiostrontium is one of the causes of cancer. To find out how much of this poison had already been disseminated throughout the world by the tests, Admiral Strauss sent a special investigation committee to all five continents and examined samples of the precipitation of strontium in plants, animals and human beings. The committee bore from its inception the cheerful cover name of "Operation Sunshine" and thoroughly lived up to its name. Its report beamed with purposeful optimism.

Of the twenty-three Japanese fishermen who had been infected, one, Kuboyama, the wireless operator, died a few months later. His countrymen call him the "first martyr of the H-bomb."

The others are still under treatment in Japanese hospitals. One of them, a fisherman named Misaki, sent the following message to the world through a reporter, Hilmar Pabel, who had visited him: "Our fate menaces all mankind. Tell that to those who are responsible. God grant that they may listen."

Nineteen

The Fall of Oppenheimer (1952-1954)

During the new phase of the armaments race, proceeding under the auspices of electronic brains, H-bombs and guided missiles, Robert Oppenheimer gradually lost his influence with the American government, which had often been considerable. The process began when in July 1952 he resigned his post as Chairman of the General Advisory Commission, that very highly respected consultative offshoot of the Atomic Energy Commission. His withdrawal had become inevitable since the victory of the activists in the "scientists' civil war," as the American lecturer John Mason Brown called the quarrel of the experts about the H-bomb. After that Oppenheimer's activities in the Atomic Energy Commission were confined to those of an occasional special consultant. He retained, however, the "Q clearance," which still allowed him access to the most carefully guarded secrets of the daily advances in the development of atomic armament. In Washington Oppenheimer's advice was only seldom called for. Over a whole year he was consulted on no more than six occasions by the Atomic Energy Commission.

Among American intellectuals Oppenheimer's reputation for imaginative guidance was rising year by year. He had become their most popular guide to the newly discovered realm of the atom. Unlike most other nuclear physicists, who had returned to their specialized pursuits after the failure of their attempts to help to mold the ideas of their countrymen, Oppenheimer continued his efforts in that direction. In a series of extremely able, sometimes magnificent, speeches he tried to indicate and if possible to bridge the gulf between the men of science and his other contemporaries. These appeals to a wide public were not couched in the usual terms. His exceptional gifts of presentation enabled him to arouse sympathy in his listeners for the great adventure of modern physics. He could communicate to them the profound excitement felt by the explorers of a new scientific region. His reputation of being a man who not only understood this strange new world but also knew how to reconcile it in philosophical and vivid images with the problem of the day was so great that in 1953 he received the honor of delivering the annual Reith Lecture for the B.B.C. It was arranged that at the same time, during his visit to England, he should receive his sixth doctorate, *honoris causa,* at Oxford.

Distinctions and marks of esteem of all kinds had been regularly conferred on Oppenheimer ever since 1945 in recognition of the work he had done during the war. They could be regarded as interest on the capital of great esteem he had then acquired. Some were of great importance, like the Medal of Merit bestowed upon him by President Truman. He appeared to be fond of trophies and collected them quite indiscriminately. Thus he accepted the Wedge Award of the Georgia Hardwood Lumber Company, and consented to be appointed Father of the Year by the National Baby Institution as well as to be nominated for the Hall of Fame of the First Half of the Century by the magazine *Popular Machanics*. His cupboards were crammed with certificates of membership in foreign academies, honorary diplomas and letters of appreciation. One of his secretaries worked several hours a day on the filing and classification in his volumes of press clippings of every item of news, article, caricature or photograph referring to her chief. Fame was a glorious thing and Oppie, however ascetic his lean countenance, now so angular, might seem, obviously enjoyed his celebrity. It was only his purely scientific prestige that declined. The name of J. R. Oppenheimer, once known as that of the author of important articles, was now hardly ever seen in physics periodicals. Between 1943, the year in which he left his study for the "great world," and 1953, his scientific publications amounted in all to only five minor contributions.

Oppenheimer's official duties as a member of no less than thirty-five different government committees gradually diminished after the new administration under President Eisenhower came to power. He was able to plan to go abroad more often. In the summer of 1953 he lectured in South America. In the late autumn he left with his wife for Europe. Among the many friends whom Oppie visited on that continent was Haakon Chevalier, whose fate had never ceased to trouble his conscience.

The professor of Romance languages had not been able to find another teaching appointment in the United States after he had become politically suspect because of the exaggerated importance attributed to Oppenheimer's tales of espionage. Finally he had been obliged to leave his home and settle in Paris as a professional translator. As Chevalier had never been a Soviet agent or even in any active contact with the Communist Party, the police had not been able, despite years of investigation, to find any evidence against him. But "something always sticks," as Chevalier had to acknowledge when he applied to the American Embassy in Paris, in 1950, for a renewal of his passport. The difficulties he encountered induced him to communicate with Oppenheimer, of whose part in the "Chevalier affair" he still remained ignorant. Oppenheimer sent him a letter which he used to obtain the renewal of his passport.

In the winter of 1953, Chevalier had every reason to continue to regard Oppie as a friend and influential protector. His delight at the prospect of seeing Oppenheimer again after so many years was so overwhelming that as soon as he heard that Oppie wanted to visit him he left a Milan conference where he was an interpreter, though he needed the fees badly, and dashed back to the French capital.

The meeting between the two men, after the years of separation, was extremely cordial. The Chevaliers had prepared a regular celebration feast in their two-room apartment on the *butte* of Montmartre. Their two wives were present, and the talk mainly concerned family and mutual friends. The subject of politics was avoided. Only once that evening did they touch on a delicate matter. Chevalier had critized the execution of the two Rosenbergs, husband and wife, sentenced to death for atomic-research espionage. Oppenheimer considered that they had deserved punishment, but he deprecated the severity of the sentence. It would now have been possible for him to take that opportunity of admitting to his friend that in 1943, during his examination by Colonel Pash, he had made Chevalier—originally with the intention of withholding his name—the hero of a highly colored tale of spying. Oppenheimer could not bring himself to make the disclosure. A second and more conspicuous chance for a belated confession of this kind occurred when Chevalier remarked that he was probably going to lose his employment as an interpreter with UNESCO. He said he could not hope to get through the security grilling which had recently been ordered for all Americans working for that body. A statement by Oppenheimer of the true facts of the case would have at last afforded Chevalier an explanation of the reason for all his difficulties. It would have put him in a position to defend himself effectively against the suspicions of the authorities. But even now Oppenheimer did not speak.

Oppie embraced both his friends on taking his leave. Today Chevalier still shudders at the recollection of that parting gesture. He never saw Oppenheimer again. Nor did he ever receive from him anything but a single, exceedingly brief, formal and noncommittal reply to the letters he wrote six months later, after he had learned the whole truth.*

* Chevalier first came to know of the true situation in the following way. He had been out shopping in Montmartre with his black poodle Coquecigru. As usual the dog had carried the evening paper, *Le Monde*, for his master. As soon as the latter had returned home and sat down, the poodle presented him with the paper. "I opened it," Chevalier says, "saw Oppie's name in the sensational headline of a report from Washington, began reading and came across my own name. Then at last I began to understand what had secretly been shaping my destiny for so many years. . . ."

By the winter of 1953, so far as Oppenheimer personally was concerned, the Chevalier affair, now ten years old, seemed to be dead and buried. But it had not been forgotten by the American authorities. J. Edgar Hoover, head of the F.B.I., had never considered that this particular episode in Oppenheimer's life had been satisfactorily explained. In 1947 he had argued in vain against the grant of a security clearance to Oppenheimer. Since then, on Hoover's instructions, his agents had been busy collecting further evidence. The Washington correspondent of the New York *Herald Tribune,* Robert J. Donovan, reports that in 1953 Oppenheimer's file, if all the papers composing it were piled on top of one another, would have reached a height of four feet six inches, almost as tall as a human being.

In November 1953, while Oppenheimer was in England, Hoover composed a digest of this formidable mass of documentary material. On the last day of the month he sent this summary not only to all interested government officials but also to President Eisenhower. The immediate occasion of this renewed interest in the Oppenheimer case had been a letter from William L. Borden, formerly senior assistant to Senator McMahon. In that communication, dated November 7, 1953, he had expressed the opinion, based on his personal knowledge of the secret J.R.O. (J. R. Oppenheimer) file, that the scientist was probably a Soviet agent in disguise.

President Eisenhower usually made it a rule not to become personally involved in any of the numerous security proceedings at that time instituted against officials who were politically suspect. In this exceptional case he nevertheless ordered an urgent special meeting at the White House. It took place on December 3, 1953. Present were two members of the Cabinet—Attorney-General Brownell and Secretary of Defense Wilson; as well as Robert Cutler, a member of the National Security Council and Lewis Strauss, Chairman of the Atomic Energy Commission. After a brief discussion—Eisenhower was about to leave for the Bermuda Conference—the

President decreed the immediate erection of a blank wall between Oppenheimer and all government secrets.

In the second half of December Oppenheimer, completely unaware of the storm brewing against him, returned to Princeton to spend the holidays with his two children. An urgent telephone call from Admiral Strauss reached him there. The head of the A.E.C. insisted that he come immediately—before Christmas, in fact—to Washington.

On the afternoon of December 21 Oppenheimer entered the Admiral's office, Room 236 in the gleaming white building of the A.E.C. on Constitution Avenue. The visitor found, to his astonishment, that Strauss was not alone. His general manager, K. D. Nichols, was standing beside him. This was the same Nichols whom Oppenheimer had met for the first time eleven years before, at another turning point in his career, in the Pullman compartment where he had drawn up, with Nichols and General Groves, plans for the first atomic-armaments laboratory.

The three men took their seats at a long table used for board meetings. Although Strauss's attitude to Oppenheimer had not been very friendly for some years, the Admiral found it difficult to break the bad news to him at once. He alluded, to begin with, to the recent death of Admiral Parsons, who had played a conspicuous part in the history of atomic armament. It was Parsons who in 1945, during the flight of the *Enola Gay* with the Thin Man, the bomb destined for the unlucky city of Hiroshima, had prepared the mechanism in the dark rear cabin of the aircraft for explosion. But these reminiscences were listened to with only half an ear by the others present. Each of them was waiting for the real subject to be broached. Quite suddenly, Strauss let fly. Oppenheimer's features grew ashen. Nichols, who later wrote an account of the meeting, recalls that Oppenheimer's first reaction was to offer his immediate resignation from his post as consultant to the Atomic Energy Commission. Strauss's next act was to pass across to him Nichols's draft of a letter in which the charges laid by the Commission were specified.

Oppenheimer glanced through the document. Twenty-three paragraphs dealt with his "associations" with Communists. But the big surprise was contained in the twenty-fourth. The scientist was censured for having "strongly opposed" the construction of the hydrogen bomb not only before but also after President Truman's decision. The letter concluded by "raising questions as to your veracity, conduct and even your loyalty."

Strauss stood up. He gave Oppenheimer a day to decide whether he would resign immediately of his own accord or would prefer the matter to be handled by a loyalty board. Oppenheimer returned home and wrote the following short letter to the chairman of the Atomic Energy Commission.

DEAR LEWIS,—Yesterday, when you asked to see me, you told me for the first time that my clearance by the Atomic Energy Commission was about to be suspended. You put to me, as a possibly desirable alternative, that I request termination of my contract as a consultant to the Commission and thereby avoid an explicit consideration of the charges on which the Commission's action would otherwise be based. I was told that if I did not do this within a day I would receive a letter notifying me of the suspension of my clearance and of the charges against me, and I was shown a draft of that letter.

I have thought most earnestly of the alternative suggested. Under the circumstances this course of action would mean that I accept and concur in the view that I am not fit to serve this government, that I have now served for some twelve years. This I cannot do. If I were thus unworthy I could hardly have served our country as I have tried or been the director of our Institute in Princeton or have spoken, as on more than one occasion I have found myself speaking, in the name of our science and our country.

Since our meeting yesterday you and General Nichols told me that the charges in the letter were familiar charges and since the time was short I paged through the letter quite briefly. I shall now read it in detail and make appropriate response.

Faithfully yours,
Robert Oppenheimer

On the following day, December 23, 1953, Nichols's letter containing the charges, which the scientist had rapidly looked through in Washington, was delivered to him officially. From that moment his access to all government secrets was barred. Security officials of the Atomic Energy Commission arrived in Princeton and cleared the safe in which Oppenheimer, with the agreement of the Commission, was in the habit of storing certain documents stamped "Secret" or "Very Secret."

Oppenheimer had always shown great interest in the Dreyfus Affair. Now he must have felt like that unjustly suspected French officer whose epaulets were torn from his uniform and whose sword was broken before his eyes. He had served the United States while neglecting many moral scruples which he could never quite abandon. Had all this been in vain? Was he to be thrust back forever into no man's land?

The public heard nothing about the forthcoming proceedings against Oppenheimer until three months later, in April 1954. Then his attorney Lloyd Garrison, at the start of the proceedings which were to be held *in camera,* handed to James Reston, head of the Washington office of the New York *Times,* General Nichols's letter of accusation and Oppenheimer's forty-three-page answer, dated March 4, 1954.

The American government's action against Oppenheimer made a deep impression. It was not only because the accused scientist was a famous man who for many people had become the symbol of the atomic age, but also because almost every one of his contemporaries was personally and intensely moved by the fate of this conscience-stricken scientist. Hardly two weeks had passed since Admiral Strauss had been forced, by the worldwide anxiety aroused by the disaster to the Japanese fishermen, to announce for the first time, on his return from the testing ground in the Pacific, more precise, official information about the formidable effects of the hydrogen bombs exploded there. A number of newspapers now wrote that Oppenheimer's associations with Communists, which had long been known and lay far back in his past, had been resurrected only because he had "opposed the hydrogen bomb." Shocked and alarmed people throughout the world, who had for years not been allowed to participate in discussions of a question which affected them all so deeply, now regarded him as their champion. To the man in the street at that moment Oppenheimer was the only truly sensitive and reflective scientist among all those involved in the construction of the new weapons. Even before the proceedings had started the halo of a martyr was bestowed on him.

Oppenheimer's professional colleagues were almost unanimous, from the beginning, in taking his side. But their support of him was only very rarely due to personal sympathy. They knew the story of his vacillations and compromises since 1945 too well to consider him, as a wider public, in its ignorance of the true facts of the case, was bound to consider him, a steadfast champion of humanity. Their primary motives were professional solidarity and self-interest. If an expert who had advised the government could afterwards be called to account for views expressed in his professional capacity and threatened with ignominious dismissal, the same thing might well happen later on to any of his colleagues. Many of them had repeatedly found him, ever since his support of the May-Johnson Bill backed by the War Department, only too submissive and docile in complying with the demands of the government. That he among all of them should now be pilloried as a saboteur seemed an historical irony. Scientists like the Nobel prize winner Harold C. Urey and Edward U. Condon, former Director of the Bureau of Standards, had often criticized Oppenheimer earlier for his excessive political pliability and timidity. Now, to their own astonishment, they found themselves rushing to his defense.*

* Condon himself, some years before, had been the object of unjustified attacks made upon his loyalty. On that occasion prominent men of science had ostentatiously invited him to take the chair at a banquet. Oppenheimer was then the only prominent scientist who had declined, "on tactical grounds," to associate himself with this demonstration of protest.

Twenty

At the Bench of the Accused *(1954-1955)*

Proceedings in the case of J. Robert Oppenheimer began on April 12, 1954. They lasted three full weeks. It was emphasized at the start that this was no trial, but purely an administrative investigation. Nevertheless, certain trial procedures, such as the taking of statements from and cross-examining of witnesses, were used. Moreover, Roger Robb, who represented the Atomic Energy Commission, adopted the style of an aggressive and pitiless public prosecutor. He did not treat Oppenheimer as a witness in his own case but as a person charged with high treason.

Not a single member of the public was admitted to the proceedings. They took place in Building T-3, an unpretentious temporary wartime office building. The white planks of its façade, the wooden "bridges of sighs" that connected its sheds, and its ugly, greenish, makeshift roof were almost exactly like those of the first administrative building in Los Alamos, where Oppenheimer's office had been when he was director. So that his arrival should be unnoticed, he was always taken in through a back door and led up to Room 2022 on the second floor. It was an ordinary office, some 24 feet long and 12 feet wide, which had been turned into a kind of courtroom by setting a few tables and chairs along the walls. Along one of the end walls sat the three members of the Personnel Security Board specially appointed by the Atomic Energy Commission to preside over these proceedings. Gordon Gray, their chairman, was an intelligent, good-looking but rather colorless official. He was the son of a millionaire and had distinguished himself particularly in the public service as Undersecretary of the Army. At the time he was President of the University of North Carolina and owner of a number of newspapers and broadcasting stations. To his right sat taciturn Thomas A. Morgan, industrialist, president, until 1952, of the Sperry Gyroscope Company. On Gray's left sat Ward V. Evans, a distinguished professor of chemistry. His occasional facetious questions and his habit of nonchalantly asking the witnesses called, while they were actually under his examination, about various private or professional acquaintances created a certain relaxed atmosphere in the tragically serious proceedings.

At the other end of the room, opposite the three judges, was an old leather couch. It was there that the witnesses sat after the usual oath had been solemnly administered. No fewer than forty prominent scientists, politicians and members of the Armed Services gave evidence. Robb, the prosecutor, sat with his back to the light from the windows. Oppenheimer, with his defense counsel, sat opposite. Seldom more than ten or twelve persons were ever present at the same time, but at times disembodied voices were heard coming from a portable loudspeaker, uttering statements by Oppenheimer himself, which had been recorded without his knowledge during his wartime examinations. They were now used contrapuntally by the prosecution, to contrast with his present evidence.

Throughout the whole of the first week of the proceedings Robert Oppenheimer was continuously examined, except twice when statements were made by others, from the start of the morning session until the evening. Rarely can any man of our time have talked so much, with such readiness and in such detail, about himself, his hopes and fears, his achievements and his mistakes. It would be impossible for any written autobiography—for such works are always subjected to self-criticism, misrepresenting or censoring the writer's experiences—to compare for authenticity with the published record of the monologues and dialogues that took place at that time in Room 2022. It ran to 992 closely printed pages.

One cannot help being struck, in reading the transcript of the hearing, by the inarticulate and diffident way in which Oppenheimer, on other occasions such a brilliant speaker, holding his listeners spellbound, then expressed himself. It almost seems as though he had voluntarily renounced the use of his most powerful asset. It is only in the written statements which he prepared before the proceedings began, as in the moving though brief account of his life, that one can detect a master of language.

Eyewitnesses who had known Oppenheimer as a dominant figure in any discussion, with eloquence which won even his opponents to adopt his own views, said that at these proceedings he often gave the impression of absent-mindedness. It was reported that "he leaned back lazily, sometimes as though his thoughts were elsewhere, on the sofa which had been turned into a dock for the occasion." André Malraux, who had met Oppenheimer through Chevalier (Malraux's translator for a number of years), is said to have remarked, after reading the transcript of the proceedings, that he could not understand why so distinguished a scientist put up with often insulting treatment from his principal adversary, Roger Robb. The great French writer is reported to have exclaimed: "He ought to have stood up proudly and shouted, 'Gentlemen, *I* am the atomic bomb!' "

But that was just what Oppenheimer's character forbade him to do. There was always more in him of Shakespeare's Danish prince than of the *Roi soleil.* Like Hamlet, he had once imagined himself born to set the time (or rather the world) right. But that "open mind" of his—a phrase he often used to describe his noncommittal attitude—always caused him to hesitate, delay and vacillate so long before taking important decisions that he was driven in the end, either by private ambition or public pressure, to come to conclusions, which he almost invariably regretted shortly afterwards.

So complex and inconsistent a character as Oppenheimer's was bound to be at a disadvantage from the start in confronting a one-track mind like Roger Robb's. The public prosecutor pitilessly involved the defendant in contradictions, lured him into traps and drove him into corners. But by exposing his vacillating adversary in this fashion he was in reality rendering Oppenheimer a great service. For the "father of the atom bomb" would never, after this, be able to seem to posterity as unscrupulous or even wicked as some of his disappointed friends believed him to be. He could only now appear as a man tortured by conflicting impulses, weak and wanting in that underlying steadiness of mind which probably only some faith above and beyond reason could have given him.

Perhaps Oppenheimer's mental distress and failure to meet the situation are nowhere so apparent as in the following passages from his dialogues with Robb.

Robb: Did you oppose the dropping of the atom bomb on Hiroshima because of moral scruples?

Oppenheimer: We set forth our——

Robb: I am asking you about "I," not "we."

Oppenheimer: I set forth my anxieties and the arguments on the other side.

Robb: You mean you argued against dropping the bomb?

Oppenheimer: I set forth arguments against dropping it.

Robb: Dropping the atom bomb?

Oppenheimer: Yes. But I did not endorse them.

Robb: You mean having worked, as you put it, in your answer, rather excellently, by night and by day for three or four years to develop the atom bomb, you then argued it should not be used?

Oppenheimer: No. I didn't argue that it should not be used. I was asked to say by the Secretary of War what the views of scientists were. I gave the views against and the views for.

Robb: But you supported the dropping of the bomb on Japan, didn't you?

Oppenheimer: What do you mean, support?

Robb: You helped pick the target, didn't you?

Oppenheimer: I did my job, which was the job I was supposed to do. I was not in a policy-making position at Los Alamos. I would have done anything that I was asked to do, including making the bombs a different shape, if I had thought it was technically feasible.

Robb: You would have made the thermonuclear weapon, too, wouldn't you?
Oppenheimer: I couldn't.
Robb: I didn't ask you that, Doctor.
Oppenheimer: I would have worked on it.
Robb: If you had discovered the thermonuclear weapon at Los Alamos, you would have done so. If you could have discovered it you would have done so, wouldn't you?
Oppenheimer: Oh, yes.

On April 22, 1954, Robert Oppenheimer became fifty years old. In normal circumstances the day would have been one of rejoicing for so successful a man. But it was a day of judgment. It fell in the second week of the proceedings, during the long procession of witnesses. All who had spoken had been full of praise for Oppenheimer. They had commended his energy and the qualities of leadership he had shown as director at Los Alamos, his realization of the need for strict measures against espionage, his organizing capabilities and his loyalty.

Before each witness, often after hours of examination by Gray, Robb and Oppenheimer's counsel, came to the end of his evidence, Professor Evans usually cut in to put a question about the personal characters and habits of the scientists who testified. He did so on this particular morning.

Bradbury, Oppenheimer's successor as Los Alamos director, was seated on the leather couch. Evans addressed him in the following terms:

Dr. Evans: Do you think that scientific men as a rule are rather peculiar individuals?
Bradbury: When did I stop beating my wife?
Mr. Gray: Especially chemistry professors?
Dr. Evans: No, physics professors.
Bradbury: Scientists are human beings . . . a scientist wants to know. He wants to know correctly and truthfully and precisely. . . . Therefore I think you are likely to find among people who have imaginative minds in the scientific field individuals who are also willing, eager, to look at a number of other fields with the

same type of interest, willingness to examine, to be convinced and without *a priori* convictions as to rightness or wrongness, that this constant or this or that curve or this or that function is fatal.

I think the same sort of willingness to explore other areas of human activity is probably characteristic. If this makes them peculiar I think it is probably a desirable peculiarity.

Dr. Evans: You didn't do that, did you?

Bradbury: Well——

Dr. Evans: Do you go fishing and things like that?

Bradbury: Yes, I have done a number of things. Some people and perhaps myself among them, I was an experimental physicist during those days, and I was very much preoccupied by the results of my own investigations.

Dr. Evans: But that didn't make you peculiar, did it?

Bradbury: This I would have to leave to others to say.

Dr. Evans: Younger people sometimes make mistakes, don't they?

Bradbury: I think this is part of people's growing up.

Dr. Evans: Do you think Dr. Oppenheimer made any mistakes?

Thus the main subject of debate, Robert Oppenheimer, had again cropped up. He sat listening with his face like a Roman mask. The day before, Rabi, a lively, sharp-tongued little Nobel prize winner, who had known Oppenheimer when he was a student and young lecturer in Europe, had uttered the following remarkable opinion of the present proceedings:

"That is what novels are about. There is a dramatic moment and the history of the man, what made him act, what he did and what sort of person he was. That is what you are really doing here. You are writing a man's life."

This statement and very many others made between April 12 and May 6, 1954, in Room 2022 characterized not only the life story of a single man but also that of a whole generation of atomic scientists. The proceedings revealed their untroubled youth, their dread of the dictators, how they were dazzled by the overwhelming nature of their discoveries, the heavy responsibility for which they had not been prepared, the fame which threatened to be their ruin, their inextricable involvement and their deep distress. Not only Robert Oppenheimer's fate was being discussed in that narrow courtroom. The debate concerned all the new, unsolved problems with which the onset of the atomic age had confronted scientists. It concerned the new part they had to play in society, their uneasiness in a world of mechanized terror and counterterror which they themselves had helped to create, above all their loss of that deeply rooted set of ethical beliefs out of which all science had formerly grown.

To study the statements made by these outstanding minds at this hearing, statements about themselves and their fate which at the time they never dreamed would be laid before a wider public, is to ask oneself why such first-rate calculators found such an entirely unexpected answer to the calculations they had made for their own lives. How paradoxical had their destiny been! They, who had been drawn into the storm center of politics, had taken up their calling in the first place mainly because they wished to turn their backs upon a chaotic and lawless world. How had it happened that men who had tried to find a more comprehensive truth were in the end obliged to spend the best years of their lives in the search for more and more perfect means of destruction?

Some statements made at these proceedings prove that all this had been too much for even the most loyal citizens. This is evident, for example, from a crisis in the life of James Conant, described not by himself, though he was one of the witnesses. During a drive from Berkeley to San Francisco in the summer of 1949, says Luis Alvarez, "Dr. Lawrence was trying to get a reaction from Dr. Conant on the possibility of radiological warfare and Dr. Conant said he wasn't interested. He didn't want to be bothered with it. I have the strong recollection that Dr. Conant said something to the effect that he was getting too old and too tired to be an adviser on affairs of this sort. He said, 'I did my job during the war' and intimated that he was burned out and he could not get any enthusiasm for new projects."

It is staggering to hear that Alvarez did not regard this final insurmountable disgust before the fact of what had become the perversion of science as an expression of deep moral principles but merely as an indication that the great scholar and teacher was simply "old, tired and burned out."

The questions put by Dr. Evans during these proceedings were often supposed to be minor and off the main point of the Oppenheimer case. But in reality it was these very queries which were aimed, for all their assumed naïveté, at the very heart of the biggest problem of all, which Oppenheimer's fall had thrown into relief: What was the true character of this new figure, the scientist, so powerful and yet so powerless?

This unconventional chemistry professor, invariably humane even when acting as a member of an official investigation committee, probably spoke for all those ordinary citizens who, he may have felt, regarded him with a mixture of admiration and fear since they had ceased to consider his profession as comic and now looked upon it as something terrible. "Are scientists peculiar people?" The crude queries of this sort, put by Evans, were in reality echoes of the voices of millons of anxious persons who would have liked the men of science to answer such questions as: "Are you the same sort of beings as we are? Do you still see any sense in moderation, in the dignity of man and the commands of his Creator? Won't you tell us what you are really after?"

The atomic scientists who gave evidence in turn before the Personnel Security Board were themselves, in fact, also before the bar of justice. And the critical question which they ought to have answered was not "Have you been loyal to the state?" but "Have you been true to mankind?"

The last act—for the time being—of the Oppenheimer drama is reminiscent, in its simplicity, of the popular ballads and traditional spectacles of earlier centuries, in which Marlowe and Goethe discovered the materials for their tragedies on the theme of Faust. Oppenheimer's fall was officially recognized by the decision of the Personnel Committee, in which Gray and Morgan voted against Evans's vote for reinstatement of the security clearance. This decision was finally ratified by the subsequent rejection of Oppenheimer's appeal by the Atomic

Energy Commission. The Commission cast four votes for rejection of the appeal against only one—Henry D. Smyth's—for its acceptance.*

Yet from that moment the subject of this ordeal began a new ascent to different heights with a purer atmosphere. Freed from the burden of official duties and the obligation to offer political and strategic advice, Oppenheimer devoted himself mainly to directing the work of the Institute for Advanced Study and investigating the spiritual and intellectual problems raised by modern nuclear physics. Those who meet him today notice traces of inward conflicts and defeats on his features,

* The Personnel Committee had stated: "In arriving at our recommendation we have sought to address ourselves to the whole question before us and not to consider the problem as a fragmented one either in terms of specific criteria or in terms of any period in Dr. Oppenheimer's life, or to consider loyalty, character and associations separately.

"However, of course the most serious finding which this Board could make as a result of these proceedings would be that of disloyalty on the part of Dr. Oppenheimer to his country. For that reason we have given particular attention to the question of his loyalty and we have come to a clear conclusion, which should be reassuring to the people of this country, that he is a loyal citizen. If this were the only consideration, therefore, we would recommend that the reinstatement of his clearance would not be a danger to the common defense and security.

"We have, however, been unable to arrive at the conclusion that it would be clearly consistent with the security interests of the United States to reinstate Dr. Oppenheimer's clearance and therefore do not so recommend.

"The following considerations have been controlling in leading us to our conclusion:

1. We find that Dr. Oppenheimer's continuing conduct and associations have reflected a serious disregard for the requirements of the security system.

2. We have found a susceptibility to influence which could have serious implications for the security interests of the country.

3. We find his conduct in the hydrogen bomb program sufficiently disturbing as to raise a doubt as to whether his future participation, if characterized by the same attitudes in a government program relating to the national defense, would be clearly consistent with the best interests of security.

4. We have regretfully concluded that Dr. Oppenheimer has been less than candid in several instances in his testimony before this Board.

Respectfully submitted.

Gordon Gray, Chairman.
Thomas A. Morgan."

which have greatly altered. At the same time they can perceive a tranquillity in his face, gained from stoical humility. His main aim seems now to be to collaborate in elucidating the grave questions which the character of our age and its unlimited technical powers have placed before us.

Edward Teller, on the contrary, the only eminent scientist who spoke against Oppenheimer and thus made a decisive contribution to the eclipse of his rival, gives today the impression of a much-troubled man, intensely uneasy about his loudly proclaimed reputation as the father of the hydrogen bomb. During the first months after the Oppenheimer hearing Teller was treated like a leper by his professional colleagues, or, even worse, as a government informer, in whose presence it was impossible to speak frankly. He insisted on being allowed to defend his position before a gathering of his associates at Los Alamos. He was listened to in an icy silence, but convinced no one.

Concealing his helplessness under an outward show of aggressive cockiness, Teller turned for advice to Enrico Fermi, one of the few people to whose opinions he was always ready to give way.

The interview between the two men took place under unusual circumstances. Fermi was in bed. He had known for some weeks that he was suffering from cancer and had little hope of recovery. The fact was no secret to Teller either. It encouraged him to speak more openly than he had ever dared to do before. "One usually reads," he remarked in recalling the occasion, "that dying men confess their sins to the living. It has always seemed to me that it would be much more logical the other way about. So I confessed my sins to Fermi. None but he, apart from the Deity, if there is one, knows what I then told him. And Fermi can at most have passed on the information in heaven."

One result of this conversation, conducted in the shadow of the awareness of death and human weakness, was that Teller was helped and supported by Fermi in the writing of an article for the periodical *Science*. The article, convincing in its personal modesty and sincerity, describes the development of the hydrogen bomb. It was entitled "The Work of Many People" and refuted the widespread popular story, not altogether unassisted in its currency, until then, by Teller himself, to the effect that he had discovered and produced the Super almost alone, in the face of resistance from his colleagues at Los Alamos. The article did not go unrecognized. Teller was again received into the community of atomic scientists. He was no longer avoided. He was tolerated. But even then he was never really forgiven.

The grounds for the animosity against him probably lie deeper than most physicists themselves realize. Teller is not only regarded as having betrayed one of his professional colleagues, but as the living example and the embodiment of a traitor to the ideals of science. As he raced from his lectures to his bomb laboratory at Livermore, as he flew from the classroom to conferences with the State Department or the Strategic Air Command, he had grown to be a vivid symbol of the restlessness and captivity of science itself.

His interviews and public lectures, even his purely scientific papers, which are considered by his colleagues to be for the most part insufficiently thought out and consequently imprecise, seem often to be contrived with a view to obtaining as many newspaper headlines in as thick type as possible. His flair for sensational news leads him sometimes to proclaim the vision of a world of moles in which a part of humanity, buried under cement and concrete, is to await the end of a third world war, and sometimes to prophesy the "happiest of all centuries." An atomic physicist still in close contact with Teller observes: "It's a pity he has no time nowadays for serious work. We are losing in him a greatly creative spirit. A short time ago I told Edward's wife that I should like to write something with him again. But she begged me not to mention any such project to him, as he already had hardly any time left for his own affairs and his family. So I never said a word about it."

When Stan Ulam, Teller's former associate, is asked about him, he cautiously refrains from committing himself to a personal opinion. Instead of doing so he takes down a book by

Anatole France from the shelf and points to a quotation heading one of the chapters. It reads: "Did you not see that they were angels?"

It is not clear whether he regards Teller as a good or a bad angel. He doesn't say but only smiles. He may think, as many atomic scientists think today, that Teller, because he surpassed everyone else in advocating, taking part in and carrying to extreme lengths the mania for armaments, acted as the instrument of a divine will and helped in the establishment of peace.

Oppenheimer, on the other hand, at present still considers the years of his blindness and distress as a part of history. "We did the devil's work," he told a visitor in the early summer of 1956, by way of summing up his experiences. "But we are now going back to our real jobs. Rabi for instance was telling me only the other day that he intended to devote himself exclusively to research in the future."

Nevertheless, many of those who once knew Oppenheimer well and were disappointed in him, cannot believe that he has renounced power forever. One of Oppie's former pupils, himself today a distinguished scientist, comments skeptically: "I'm afraid he has only assumed a new role in his big repertoire. Just now he happens to be, of necessity, saint and martyr, but if ever the wind changes, he'll be busy again in Washington with the rest of them."

This bitter judgment is understandable enough, no doubt, but as a prophecy it is scarcely reliable. More important and stimulating tasks await Oppenheimer than any which the Cabinet or General Staff Corps could offer him.

In one of his most recent lectures he indicated imaginatively the goal to which he wished to dedicate himself: "Both the man of science and the man of art live always at the edge of mystery, surrounded by it. Both, as the measure of their creation, have always had to do with the harmonization of what is new with what is familiar, with the balance between novelty and synthesis, with the struggle to make partial order in total chaos. They can, in their work and in their lives, help themselves, help one another, and help all men. They can make the paths which connect the villages of art and science with each other and with the world at large into the multiple, varied and precious bonds of a true and world-wide community. This cannot be an easy life. We shall have a rugged time of it to keep our minds open and to keep them deep, to keep our sense of beauty and our ability to make it, and our occasional ability to see it in places remote, strange and unfamiliar. We shall have a rugged time of it, all of us, in keeping these flourishing in a great, open, windy world. And in this condition we can help, because we can love, one another."

Epilogue: The Last Chance?

Today Room 2022, where the inquiry into the Oppenheimer case was held, once more serves as an ordinary office. It is occupied by a civilian employee of the Navy who does not even know what went on between those four walls four years ago. At Farm Hall, where the German nuclear physicists were once interned, the new owner produces still-life paintings of flowers. At Haigerloch, in the former underground laboratory, rabbits peacefully munch their hay. The crater plowed up by the first atom bomb in the barren soil of New Mexico has long since been filled in.

And still the atomic scientists' unrest has not disappeared. It has grown with the problem. "What should we do?" asked C. F. von Weizsäcker in the fall of 1945. "We have played with fire like children, and it flared up before we expected it." The questions of conscience with which almost every nuclear physicist had confronted himself since the end of the war have found no acknowledged and binding answers even until today.

Atomic scientists are living more tranquil lives since the summer of 1955, when many of them met at Geneva at the conference on "Atoms for Peace." Security regulations have been slightly relaxed. The obstacles to the resumption of international scientific communications have been removed to some extent. It is true that improved hydrogen bombs have continued to explode on the testing grounds in the Southwest Pacific and in Soviet Asia, unperturbed by manifestoes and resolutions. But physicists in general today are more deeply concerned with the long neglected investigation of principles and with the problems of economic exploitation of nuclear energy than they are with those of weapons research and development.

Unlike the characters in plays and novels, who vanish when the curtain falls or the last pages are turned, the heroes of history often survive the end of their tragedy. A reporter finds them busy with their everyday work, full of new plans and fresh hopes, their eyes more on what is coming than on what is past.

No doubt most natural scientists now recognize that they share the responsibility for the use made of their discovery. Some of them believe it means that they should not participate in weapons projects. A landmark of that resolute attitude has been the declaration of the eighteen German atomic scientists led by Max Born, Otto Hahn and Weizsäcker, issued on April 12, 1957. It ends with this statement: "At all events not one of the undersigned is prepared himself to share in the production, the testing, or the stock-piling of atomic weapons in any manner."

Others, on the contrary, think that their newly acquired sense of responsibility forces them to take part in the weapons programs of their nations whether they like it or not. Edward Teller has quite recently outlined this position in his testimony before the Senate Foreign Relations Subcommittee on Disarmament with the following outcry: "I chose the profession of a scientist and I am in love with science; and I would not do willingly or eagerly anything else but pure science because it is beautiful and my interest is there. I don't like weapons. I would like to have peace. But for peace we need weapons and I do not think my views are distorted. I believe I am contributing to a peaceful world."

A growing number of scientists still aim at international control of atomic research by means of reciprocal inspection. Others consider such a system impossible in the present state of nuclear technology, which has put it out of date. Some atomic scientists, disappointed by their failure to influence public life, have withdrawn to their laboratories. Some advocate an even more intensive preoccupation with the outside world. Of the younger men quite a number regard their scientific work as a kind of intellectual competition, not involving any particularly deep meaning or obligation. But again, even some of these find religious experience in their research.*

The most significant effort towards re-establishing the international family of scientists and giving it a voice in the conduct of human affairs has been the annual meetings, which have become known as the "Pugwash Conferences."

This movement, like so many other events in the political history of atomic development, received its decisive momentum from the late Albert Einstein. Two days before his death he put his signature on a statement about the nature of future wars, drafted with his approval by the eminent philosopher and mathematician Bertrand Russell. This may have been the last public act of the discoverer of relativity.

Russell wanted to submit this statement to an international body of scientists and persuaded the British Parliamentary Group for World Government (a group of Members of Parliament founded by the Laborite Henry Usborne) to convene such a conference through the World Association of Parliamentarians for World Government.

This meeting, held in the Chambers of the London County Council from August 3 to 5, 1955, did not at first look as if it might be of much consequence. Eugene Rabinowitch remembers: "Apart from the preparation of Bertand Russell's statement, the program of the London conference was rather improvised, and no personal invitations were extended to scientists most likely to contribute to it—one reason being that the actual organization of the conference was in the hands of people unfamiliar with the world of science, and the other, that practically no funds were available for travel expenses. Invitations were sent to the rectors or presidents of all universities in the world, with the request to transmit them to interested faculty members. It was hoped that

* See Richard P. Feynman's "The Relation of Science and Religion" in *Engineering and Science* for June 1956.

atomic physicists on their way to Geneva would stop in London; but Professor Marcus Oliphant of Australia proved to be the only prominent atomic physicist from outside England who availed himself of this opportunity. My own attempt to inform the Federation of American Scientists of the conference and to induce some individual American atomic scientists who were going to Geneva to stop over in London came too late to influence already fixed travel plans."

The sudden and unexpected appearance of a four-man Soviet delegation headed by A. V. Topchiev, Permanent Secretary of the Soviet Academy of Sciences, seems to have not only surprised, but also disconcerted, the organizers of the conference. They probably expected that the Russians, who until then had called Russell "a capitalist lackey and bloodthirsty warmonger," would merely try to torpedo the meeting or turn it into a propaganda circus. This was prevented by the firmness of the conference leadership, which stuck to its intention to have the topics of the agenda* discussed in a spirit of honest and thorough study.

From this rambling, ill-prepared, and rather unpromising meeting came, in July 1957, the first successful conference in the quaint old clipper-building town of Pugwash, situated on the Northumberland Straits of Nova Scotia. This quiet and old-fashioned place, a rather unlikely spot for a meeting of scientists concerned with the latest and most urgent problems of humanity, had been chosen in deference to the Canadian-born financier Cyrus Eaton, whose ancestors had worked and lived there. In this astonishing man the atomic scientists who refused any organizational support had found a Maecenas willing to shoulder the considerable financial responsibilities of this and later meetings while leaving them free to do what they wanted.

The first and the following two Pugwash Conferences (April 1957 at Lake Beauport near Quebec and September 1958 at

* The four topics were considered by four commissions: 1. Destructive potential of nuclear weapons. 2. Risks involved in nonmilitary atomic developments. 3. Technical possibilities of controlling atomic disarmament. 4. The responsibility of scientists.

Kitzbuehel, Austria) have deliberately limited the number of attendants and restricted the publicity. Both factors have encouraged intimate, thorough, and soul-searching conversations free from any play for public attention and—perhaps most important—the intellectual cross-fertilization caused by communication among natural scientists from different branches of science and men like a specialist of international law, a former international civil servant, an analyst of modern strategic problems, an outstanding philosopher.

Thus Pugwash does more than bridge the gulf between East and West; it tries to narrow down the growing gaps caused by specialization, and thus forms—probably without intentionally trying to do so—part of the trend towards a new universalism which aims at a "whole man."

A fat two-inch-thick volume resulted from the second Pugwash Conference. It contained the main papers contributed by the participants as well as the minutes of the discussions. This big bundle of mimeographed paper was sent to the heads of interested governments. Further publication was not contemplated. This may seem a meager result to the many people who expect intellectual and spiritual enlightenment to work as fast and efficiently as the installations controlled by switches which light our homes. But actually, since 1945 the scientists' movement, though apparently inefficient and at best of meager success, has had, in more indirect and diffuse ways, an immense and ever-growing influence upon the public mind. Its ideas, which have steadily matured, and even the terms it was often the first to coin have unconsciously and without acknowledgment become part of our public and governmental thinking.

One has only to ask oneself: What would have happened had the atomic scientists chosen to remain indifferent and silent after Hiroshima or if they had even been proud of their achievement there? Their contemporaries would then probably have been left ignorant about the nature of the nuclear revolution and the unheard-of new dangers this "quantum jump" of technology entailed for mankind. The men in power on both sides, unhampered by public opinion, would then probably have fallen prey more easily to the temptation to use their atomic swords to slash entangled political knots. By a curious detour public opinion, fired by the repeated warnings of scientific authorities, even worked on the other side of the iron curtain. Only when they tried to take the lead of the popular movement against the use of atomic weapons in the free world were the Soviet rulers finally compelled to tell the frightening facts of nuclear warfare to their own people.

This awakening toward new responsibilities has finally had definite effect upon the scientists themselves. This observer, who is quite conscious that his judgment may be premature and too subjective, believes that the intellectual uneasiness and psychological distress he has found among atomic scientists is itself a noteworthy phenomenon. For three hundred years the natural scientist believed that he could isolate himself from the world, but now he is beginning to regard himself as a part of it. He feels himself to be conditioned and limited. This realization has shown him the way to a new modesty. He has been obliged to recognize that he, like everyone else, in Bohr's words, "is both a spectator and an actor in the great drama of being."

Modern science had been inspired by "the proud will to master nature." It is an attitude which found expression, above all, in Bacon's aphorism "Knowledge is power." But today one far more often hears it in the form, "Knowledge is unfortunately power." The scientist has come to "fear his godlike character," as Feynman puts it, and confesses his "intellectual humility in the face of unanswerable secrets of the universe, which ought to remain unanswered." The age which culminated in the development of absolute weapons identified progress, almost unanimously, with progress in science and technology. But today the outstanding physicist Heisenberg declares: "The space in which man has developed as an intellectual being has more dimensions than that of the single direction in which he has moved during the last few centuries."

This new modesty, like the inhuman and superhuman weapons, has grown from the tree of atomic research. It was the study of the atomic world that taught the theoretical physicist to recognize a truth long ago proclaimed by religion but now also susceptible of scientific proof—that human capacity for observation and judgment has its limits. The atom bomb, whose power is the clearest expression of the lack of moderation in modern man, comes from the same root as the new philosophy of moderation, inspired by the experience of nuclear research.

When H. G. Wells, in 1946, shortly before his death, found that the belief in scientific progress had collapsed, he felt justified in announcing that man was "at the end of his tether" and almost inevitably doomed to rapid extinction. But probably man has only come to the end of *one* tether.

Wolfgang Pauli, formerly known to the family of atomic scientists as a skeptic, has indicated a possible road for humanity to take. At Copenhagen in 1932 Pauli had played Mephisto to Faust. But by 1955 his keen mind had so extended its field of vision that he became the eloquent exponent of a long-neglected inner way to salvation. At the close of a lecture on "Science and Western Thought" he said: "Since the seventeenth century the activities of the human spirit have been strictly classified in separate compartments. But in my view the attempt to eliminate such distinctions by a combination of rational understanding and the mystical experience of unity obeys the explicit or implicit imperative of our own contemporary age."

Can the "new modesty," along with a fresh recognition of an inner way to salvation, exert as strong an influence on the coming centuries as the spirit of overweening pride, now revealed to have been disastrous?

The author of this chronicle does not venture to prophesy. He intends only to present a picture, hoping thereby to contribute something to the great debate which may perhaps eventually lead to plans for a future without fear.

List of Sources

PERIODICALS

Bulletin of the Atomic Scientists, 1946–58 (Chicago)
Atomic Scientists News, 1950–2 (London)
Atomic Scientists Journal, 1953–6 (London)
Nature, 1939–56 (London)
Newsletter of the "Federation of American Scientists," 1946–56 (Washington)
Newsletter of the "Society for the Social Responsibility of Science," 1950–6 (Gambier, Ohio)
Die Naturwissenschaften, 1933–9 (Berlin)
Die Naturwissenschaften, 1946–56 (Göttingen)
Science, 1939, 1945–56 (Washington)
La Nef (Paris), "L'atome, notre destin" (September 1955)
Politics (New York): "The Bomb" (September 1945)
Fortune (May 1956)
Die Zeit (Hamburg): "Der deutsche Forscher-Anteil" by K. Diebner (August 1955)
"Safety Planning of an Atomic Test Operation" by Roy Reider (Transactions of the National Safety Council, 1954)

BOOKS

H. Hartmann: *Schöpfer des neuen Weltbildes* (Bonn, 1952)
J. R. Oppenheimer: *Science and the Common Understanding* (New York, 1954)
J. R. Oppenheimer: *The Open Mind* (New York, 1955)
H. Schwartz and W. Spengler: *Forscher und Wissenschaftler im heutigen Europa* (Oldenburg, 1955)

J. G. Crowther: *British Scientists of the Twentieth Century* (London, 1952)
J. Bergier and P. de Latil: *Quinze hommes et un secret* (Paris, 1955)
H. De Wolf Smyth: *Atomic Energy for Military Purposes* (Washington, 1945)
Crowther and Whiddington: *Science at War* (London, 1947)
L. Bertin: *Atom Harvest* (London, 1955)
S. A. Goudsmit: *Alsos* (New York, 1947)
S. Werner: *Niels Bohr* (Copenhagen, 1955)
M. Rouze: *F. Joliot-Curie* (Paris, 1950)
Carl Selig: *Helle Zeit—Dunkle Zeit—In Memoriam Albert Einstein* (Zurich, 1956)
A. Schilp: *Albert Einstein, Philosopher-Scientist* (New York, 1951)
A. Vallentin: *Das Drama Albert Einsteins* (Stuttgart, 1954)
L. Fermi: *Atoms in the Family* (Chicago, 1954)
A. Moorehead: *The Traitors* (London, 1952)
U.S. Atomic Energy Commission: *In the Matter of J. Robert Oppenheimer* (Washington, 1954)
J. and S. Alsop: *We Accuse—The Story of the Miscarriage of American Justice in the Case of J. Robert Oppenheimer* (New York, 1954)
A. S. Eve: *Rutherford* (Oxford, 1939)
Iris Runge: *Carl Runge und sein wissenschaftliches Werk* (Göttingen, 1949)
E. Rabinowitch: *Minutes to Midnight* (Chicago, 1950)
A. Amrine: *Secret* (Boston, 1950)
B. Barber: *Science and Social Order* (Glencoe, 1952)
R. C. J. Butow: *Japan's Decision to Surrender* (Stanford, 1954)
P. M. S. Blackett: *Military and Political Consequences of Atomic Energy* (London, 1952)
Walter Gellhorn: *Security, Loyalty and Science* (Ithaca, 1950)
J. R. Shepley and C. Blair: *The Hydrogen Bomb* (New York, 1954)
M. J. Ruggles and A. Kramish: *Soviet Atomic Policy* (Rand Corporation, Santa Monica, 1956. Duplicated)

Appendix A

Niels Bohr's Memorandum to President Roosevelt, July 1944*

It certainly surpasses the imagination of anyone to survey the consequences of the project in years to come, where, in the long run, the enormous energy sources which will be available may be expected to revolutionize industry and transport. The fact of immediate preponderance is, however, that a weapon of an unparalleled power is being created which will completely change all future conditions of warfare.

Quite apart from the question of how soon the weapon will be ready for use and what role it may play in the present war, this situation raises a number of problems which call for most urgent attention. Unless, indeed, some agreement about the control of the use of the new active materials can be obtained in due time, any temporary advantage, however great, may be outweighed by a perpetual menace to human security.

Ever since the possibilities of releasing atomic energy on a vast scale came in sight, much thought has naturally been given to the question of control, but the further the exploration of the scientific problems concerned is proceeding, the clearer it becomes that no kind of customary measures will suffice for this purpose, and that the terrifying prospect of a future competition between nations about a weapon of such formidable character can only be avoided through a universal agreement in true confidence.

* See p. 173.

In this connection it is particularly significant that the enterprise, immense as it is, has still proved far smaller than might have been anticipated, and that the progress of the work has continually revealed new possibilities for facilitating the production of the active materials and of intensifying their efforts.

The prevention of a competition prepared in secrecy will therefore demand such concessions regarding exchange of information and openness about industrial efforts, including military preparations, as would hardly be conceivable unless all partners were assured of a compensating guarantee of common security against dangers of unprecedented acuteness.

The establishment of effective control measures will of course involve intricate technical and administrative problems, but the main point of the argument is that the accomplishment of the project would not only seem to necessitate but should also, due to the urgency of mutual confidence, facilitate a new approach to the problems of international relationship.

The present moment where almost all nations are entangled in a deadly struggle for freedom and humanity might, at first sight, seem most unsuited for any committing arrangement concerning the project. Not only have the aggressive powers still great military strength, although their original plans of world domination have been frustrated and it seems certain that they must ultimately surrender, but even when this happens, the nations united against aggression may face grave causes of disagreement due to conflicting attitudes toward social and economic problems.

A closer consideration, however, would indicate that the potentialities of the project as a means of inspiring confidence under these very circumstances acquire real importance. Moreover, the present situation affords unique possibilities which might be forfeited by a postponement awaiting the further development of the war situation and the final completion of the new weapon. . . .

In view of these eventualities the present situation appears to offer a most favorable opportunity for an early initiative from the side which by good fortune has achieved a lead in the efforts of mastering mighty forces of nature hitherto beyond human reach.

Without impeding the immediate military objectives, an initiative, aiming at forestalling a fateful competition, should serve to uproot any cause of distrust between the powers on whose harmonious collaboration the fate of coming generations will depend.

Indeed, it would appear that only when the question is raised among the united nations as to what concessions the various powers are prepared to make as their contribution to an adequate control arrangement, will it be possible for any one of the partners to assure himself of the sincerity of the intentions of the others.

Of course, the responsible statesmen alone can have insight as to the actual political possibilities. It would, however, seem most fortunate that the expectations for a future harmonious international co-operation, which have found unanimous expressions from all sides within the united nations, so remarkably correspond to the unique opportunities which, unknown to the public, have been created by the advancement of science.

Many reasons, indeed, would seem to justify the conviction that an approach with the object of establishing common security from ominous menaces, without excluding any nation from participating in the promising industrial development which the accomplishment of the project entails, will be welcomed, and be met with loyal co-operation in the enforcement of the necessary far-reaching control measures.

It is in such respects that helpful support may perhaps be afforded by the world-wide scientific collaboration which for years has embodied such bright promises for common human striving. Personal connections between scientists of different nations might even offer means of establishing preliminary and unofficial contact.

It need hardly be added that any such remark or suggestion

implies no underrating of the difficulty and delicacy of the steps to be taken by the statesmen in order to obtain an arrangement satisfactory to all concerned, but aims only at pointing to some aspects of the situation which might facilitate endeavours to turn the project to the lasting benefit of the common cause.

Appendix B

The "Franck Report"

A REPORT TO THE SECRETARY OF WAR, JUNE 1945*

I. Preamble

The only reason to treat nuclear power differently from all other developments in the field of physics is the possibility of its use as a means of political pressure in peace and sudden destruction in war. All present plans for the organization of research, scientific and industrial development, and publication in the field of nucleonics are conditioned by the political and military climate in which one expects those plans to be carried out. Therefore, in making suggestions for the postwar organization of nucleonics, a discussion of political problems cannot be avoided. The scientists on this Project do not presume to speak authoritatively on problems of national and international policy. However, we found ourselves, by the force of events, during the last five years, in the position of a small group of citizens cognizant of a grave danger for the safety of this country as well as for the future of all the other nations, of which the rest of mankind is unaware. We therefore feel it is our duty to urge that the political problems, arising from the mastering of nuclear power, be recognized in all their gravity, and that appropriate steps be taken for their study and the preparation of necessary decisions. We hope that the creation

* See p. 184.

of the Committee by the Secretary of War to deal with all aspects of nucleonics, indicates that these implications have been recognized by the government. We believe that our acquaintance with the scientific elements of the situation and prolonged preoccupation with its world-wide political implications, imposes on us the obligation to offer to the Committee some suggestions as to the possible solution of these grave problems.

Scientists have often before been accused of providing new weapons for the mutual destruction of nations, instead of improving their well-being. It is undoubtedly true that the discovery of flying, for example, has so far brought much more misery than enjoyment and profit to humanity. However, in the past, scientists could disclaim direct responsibility for the use to which mankind had put their disinterested discoveries. We feel compelled to take a more active stand now because the success which we have achieved in the development of nuclear power is fraught with infinitely greater dangers than were all the inventions of the past. All of us, familiar with the present state of nucleonics, live with the vision before our eyes of sudden destruction visited on our own country, of a Pearl Harbor disaster repeated in thousand-fold magnification in every one of our major cities.

In the past, science has often been able to provide also new methods of protection against new weapons of aggression it made possible, but it cannot promise such efficient protection against the destructive use of nuclear power. This protection can come only from the political organization of the world. Among all the arguments calling for an efficient international organization for peace, the existence of nuclear weapons is the most compelling one. In the absence of an international authority which would make all resort to force in international conflicts impossible, nations could still be diverted from a path which must lead to total mutual destruction, by a specific international agreement barring a nuclear armaments race.

II. Prospects of Armaments Race

It could be suggested that the danger of destruction by nuclear weapons can be avoided—at least as far as this country is concerned—either by keeping our discoveries secret for an indefinite time, or else by developing our nuclear armaments at such a pace that no other nations would think of attacking us from fear of overwhelming retaliation.

The answer to the first suggestion is that although we undoubtedly are at present ahead of the rest of the world in this field, the fundamental facts of nuclear power are a subject of common knowledge. British scientists know as much as we do about the basic wartime progress of nucleonics—if not of the specific processes used in our engineering developments—and the role which French nuclear physicists have played in the pre-war development of this field, plus their occasional contact with our Projects, will enable them to catch up rapidly, at least as far as basic scientific discoveries are concerned. German scientists, in whose discoveries the whole development of this field originated, apparently did not develop it during the war to the same extent to which this has been done in America: but to the last day of the European war, we were living in constant apprehension as to their possible achievements. The certainty that German scientists were working on this weapon and that their government would certainly have no scruples against using it when available, was the main motivation of the initiative which American scientists took in urging the development of nuclear power for military purposes on a large scale in this country. In Russia, too, the basic facts and implications of nuclear power were well understood in 1940, and the experience of Russian scientists in nuclear research is entirely sufficient to enable them to retrace our steps within a few years, even if we should make every attempt to conceal them. Even if we can retain our leadership in basic knowledge of nucleonics for a certain time by maintaining secrecy as to all results achieved on this and associated Projects, it would be foolish to hope that this can protect us for more than a few years.

It may be asked whether we cannot prevent the development of military nucleonics in other countries by a monopoly on the raw materials of nuclear power. The answer is that even though the largest now known deposits of uranium ores are under the control of powers which belong to the "western" group (Canada, Belgium and British India), the old deposits in Czechoslovakia are outside this sphere. Russia is known to be mining radium on its own territory; and even if we do not know the size of the deposits discovered so far in the USSR, the probability that no large reserves of uranium will be found in a country which covers one-fifth of the land area of the earth (and whose sphere of influence takes in additional territory) is too small to serve as a basis for security. Thus, we cannot hope to avoid a nuclear armament race either by keeping secret from the competing nations the basic scientific facts of nuclear power or by cornering the raw materials required for such a race.

We now consider the second of the two suggestions made at the beginning of this section, and ask whether we could not feel ourselves safe in a race of nuclear armaments by virtue of our greater industrial potential, including greater diffusion of scientific and technical knowledge, greater volume and efficiency of our skilled labor crops, and greater experience of our management—all the factors whose importance has been so strikingly demonstrated in the conversion of this country into an arsenal of the Allied Nations in the present war. The answer is that all that these advantages can give us is the accumulation of a larger number of bigger and better atomic bombs.

However, such a quantitative advantage in reserves of bottled destructive power will not make us safe from sudden attack. Just because a potential enemy will be afraid of being "outnumbered and outgunned," the temptation for him may be overwhelming to attempt a sudden unprovoked blow—particularly if he should suspect us of harboring aggressive intentions against his security or his sphere of influence. In no other type of warfare does the advantage lie so heavily with the aggressor. He can place his "infernal machines" in advance in all our major cities and explode them simultaneously, thus destroying a major part of our industry and a large part of our population, aggregated in densely populated metropolitan districts. Our possibilities of retaliation—even if retaliation should be considered adequate compensation for the loss of millions of lives and destruction of our largest cities—will be greatly handicapped because we must rely on aerial transportation of the bombs, and also because we may have to deal with an enemy whose industry and population are dispersed over a large territory.

In fact, if the race for nuclear armaments is allowed to develop, the only apparent way in which our country can be protected from the paralyzing effects of a sudden attack is by dispersal of those industries which are essential for our war effort and dispersal of the populations of our major metropolitan cities. As long as nuclear bombs remain scarce (i.e. as long as uranium remains the only basic material for their fabrication), efficient dispersal of our industry and the scattering of our metropolitan population will considerably decrease the temptation to attack us by nuclear weapons.

At present, it may be that atomic bombs can be detonated with an effect equal to that of 20,000 tons of TNT. One of these bombs could then destroy something like 3 square miles of an urban area. Atomic bombs containing a larger quantity of active material but still weighing less than one ton may be expected to be available within ten years which could destroy over ten square miles of a city. A nation able to assign 10 tons of atomic explosives for a sneak attack on this country, can then hope to achieve the destruction of all industry and most of the population in an area from 500 square miles upwards. If no choice of targets, with a total area of five hundred square miles of American territory, contains a large enough fraction of the nation's industry and population to make their destruction a crippling blow to the nation's war potential and its ability to defend itself, then the attack will not pay, and may not be undertaken. At present, one could easily select in this country a hundred areas of five square miles each whose simultaneous destruction would be a staggering blow to the nation. Since the area of the United States is about three million square miles, it should be possible to scatter its industrial and human resources in such a way as to leave no 500 square miles important enough to serve as a target for nuclear attack.

We are fully aware of the staggering difficulties involved in such a radical change in the social and economic structure of our nation. We felt, however, that the dilemma had to be stated, to show what kind of alternative methods of protection will have to be considered if no successful international agreement is reached. It must be pointed out that in this field we are in a less favorable position than nations which are either now more diffusely populated and whose industries are more scattered, or whose governments have unlimited power over the movement of population and the location of industrial plants.

If no efficient international agreement is achieved, the race for nuclear armaments will be on in earnest not later than the morning after our first demonstration of the existence of nuclear weapons. After this, it might take other nations three or four years to overcome our present head start, and eight or ten years to draw even with us if we continue to do intensive work in this field. This might be all the time we would have to bring about the relocation of our population and industry. Obviously, no time should be lost in inaugurating a study of this problem by experts.

III. Prospects of Agreement

The consequences of nuclear warfare, and the type of measures which would have to be taken to protect a country from total destruction by nuclear bombing, must be as abhorrent to other nations as to the United States. England, France and the smaller nations of the European continent, with their congeries of people and industries would be in a particularly desperate situation in the face of such a threat. Russia and China are the only great nations at present which could survive a nuclear attack. However, even though these countries may value human life less than the peoples of Western Europe and America, and even though Russia, in particular, has an immense space over which its vital industries could be dispersed and a government which can order this dispersion the day it is convinced that such a measure is necessary—there is no doubt that Russia, too, will shudder at the possibility of a sudden disintegration of Moscow and Leningrad, almost miraculously preserved in the present war, and of its new industrial cities in the Urals and Siberia. Therefore, only lack of mutual trust, and not lack of desire for agreement, can stand in the path of an efficient agreement for the prevention of nuclear warfare. The achievement of such an agreement will thus essentially depend on the integrity of intentions and readiness to sacrifice the necessary fraction of one's own sovereignty, by all the parties to the agreement.

One possible way to introduce nuclear weapons to the world—which may particularly appeal to those who consider nuclear bombs primarily as a secret weapon developed to help win the present war—is to use them without warning on appropriately selected objects in Japan.

Although important tactical results undoubtedly can be achieved by a sudden introduction of nuclear weapons, we nevertheless think that the question of the use of the very first available atomic bombs in the Japanese war should be weighed very carefully, not only by military authorities, but by the highest political leadership of this country.

Russia, and even allied countries which bear less mistrust of our ways and intentions, as well as neutral countries may be deeply shocked by this step. It may be very difficult to persuade the world that a nation which was capable of secretly preparing and suddenly releasing a new weapon, as indiscriminate as the rocket bomb and a thousand times more destructive, is to be trusted in its proclaimed desire of having such weapons abolished by international agreement. We have large accumulations of poison gas, but do not use them, and recent polls have shown that public opinion in this country would disapprove of such a use even if it would accelerate the winning of the Far Eastern war. It is true that some irrational element in mass psychology makes gas poisoning more revolting than blasting by explosives, even though gas warfare is in no way more "inhuman" than the war of bombs and bullets. Nevertheless, it is not at all certain that American public opinion, if it could be enlightened as to the effect of atomic explosives, would approve of our own country being the first to introduce such an indiscriminate method of wholesale destruction of civilian life.

Thus, from the "optimistic" point of view—looking forward to an international agreement on the prevention of nuclear warfare—the military advantages and the saving of American lives achieved by the sudden use of atomic bombs against Japan may be outweighed by the ensuing loss of confidence and by a wave of horror and repulsion sweeping over the rest of the world and perhaps even dividing public opinion at home.

From this point of view, a demonstration of the new weapon might best be made, before the eyes of representatives of all the United Nations, on the desert or a barren island. The best possible atmosphere for the achievement of an international agreement could be achieved if America could say to the world, "You see what sort of a weapon we had but did not use. We are ready to renounce its use in the future if other nations join us in this renunciation and agree to the establishment of an efficient control."

After such a demonstration the weapon might perhaps be used against Japan if the sanction of the United Nations (and if public opinion at home) were obtained, perhaps after a preliminary ultimatum to Japan to surrender or at least to evacuate certain regions as an alternative to their total destruction. This may sound fantastic, but in nuclear weapons we have something entirely new in order of magnitude of destructive power, and if we want to capitalize fully on the advantage their possession gives us, we must use new and imaginative methods.

It must be stressed that if one takes the pessimistic point of view and discounts the possibility of an effective international control over nuclear weapons at the present time, then the advisability of an early use of nuclear bombs against Japan becomes even more doubtful—quite independently of any humanitarian considerations. If an international agreement is not concluded immediately after the first demonstration, this will mean a flying start towards an unlimited armaments race. If this race is inevitable, we have every reason to delay its beginning as long as possible in order to increase our head start still further.

The benefit to the nation, and the saving of American lives in the future, achieved by renouncing an early demonstration of nuclear bombs and letting the other nations come into the race only reluctantly, on the basis of guesswork and without definite knowledge that the "thing does work," may far outweigh the advantages to be gained by the immediate use of the first and comparatively inefficient bombs in the war against Japan. On the other hand, it may be argued that without an early demonstration it may prove difficult to obtain adequate support for further intensive development of nucleonics in this country and that thus the time gained by the postponement of an open armaments race will not be properly used. Furthermore one may suggest that other nations are now, or will soon be, not entirely unaware of our present achievements, and that consequently the postponement of a demonstration may serve no useful purpose as far as the avoidance of an armaments race is concerned, and may only create additional mistrust, thus worsening rather than improving the chances of an ultimate accord on the international control of nuclear explosives.

Thus, if the prospects of an agreement will be considered poor in the immediate future, the pros and cons of an early revelation of our possession of nuclear weapons to the world—not only by their actual use against Japan, but also by a prearranged demonstration—must be carefully weighed by the supreme political and military leadership of the country, and the decisions should not be left to the considerations of military tactics alone.

One may point out that scientists themselves have initiated the development of this "secret weapon" and it is therefore strange that they should be reluctant to try it out on the enemy as soon as it is available. The answer to this question was given above—the compelling reason for creating this weapon with such speed was our fear that Germany had the technical skill necessary to develop such a weapon and that the German government had no moral restraints regarding its use.

Another argument which could be quoted in favor of using atomic bombs as soon as they are available is that so much taxpayers' money has been invested in these Projects that the Congress and the American public will demand a return for their money. The attitude of American public opinion, mentioned earlier, in the matter of the use of poison gas against Japan, shows that one can expect the American public to understand that it is sometimes desirable to keep a weapon in readiness for use only in extreme emergency; and as soon as the potentialities of nuclear weapons are revealed to the American people, one can be sure that they will support all attempts to make the use of such weapons impossible.

Once this is achieved, the large installations and the accumulation of explosive material at present earmarked for potential military use will become available for important peacetime developments, including power production, large engineering undertakings, and mass production of radioactive materials. In this way, the money spent on wartime development of nucleonics may become a boon for the peacetime development of national economy.

IV. Methods of International Control

We now consider the question of how an effective international control of nuclear armaments can be achieved. This is a difficult problem, but we think it soluble. It requires study by statesmen and international lawyers, and we can offer only some preliminary suggestions for such a study.

Given mutual trust and willingness on all sides to give up a certain part of their sovereign rights, by admitting international control of certain phases of national economy, the control could be exercised (alternatively or simultaneously) on two different levels.

The first and perhaps the simplest way is to ration the raw materials—primarily, the uranium ores. Production of nuclear explosives begins with the processing of large quantities of uranium in large isotope separation plants or huge production piles. The amounts of ore taken out of the ground at different locations could be controlled by resident agents of the international Control Board, and each nation could be allotted only an amount which would make large-scale separation of fissionable isotopes impossible.

Such a limitation would have the drawback of making impossible also the development of nuclear power for peacetime purposes. However, it need not prevent the production of radioactive elements on a scale sufficient to revolutionize the industrial, scientific and technical use of these materials, and would thus not eliminate the main benefits which nucleonics promises to bring to mankind.

An agreement on a higher level, involving more mutual trust and understanding, would be to allow unlimited production, but keep exact bookkeeping on the fate of each pound of uranium mined. If in this way, check is kept on the conversion of uranium and thorium ore into pure fissionable materials, the question arises as to how to prevent accumulation of large quantities of such materials in the hands of one or several nations. Accumulations of this kind could be rapidly converted into atomic bombs if a nation should break away from international control. It has been suggested that a compulsory denaturation of pure fissionable isotopes may be agreed upon—by diluting them, after production, with suitable isotopes to make them useless for military purposes, while retaining their usefulness for power engines.

One thing is clear: any international agreement on prevention of nuclear armaments must be backed by actual and efficient controls. No paper agreement can be sufficient since neither this or any other nation can stake its whole existence on trust in other nations' signatures. Every attempt to impede the international control agencies would have to be considered equivalent to denunciation of the agreement.

It hardly needs stressing that we as scientists believe that any systems of control envisaged should leave as much freedom for the peacetime development of nucleonics as is consistent with the safety of the world.

Summary

The development of nuclear power not only constitutes an important addition to the technological and military power of the United States, but also creates grave political and economic problems for the future of this country.

Nuclear bombs cannot possibly remain a "secret weapon" at the exclusive disposal of this country for more than a few years. The scientific facts on which construction is based are well known to scientists of other countries. Unless an effective international control of nuclear explosives is instituted, a race for nuclear armaments is certain to ensue following the first revelation of our possession of nuclear weapons to the world. Within ten years other countries may have nuclear bombs, each of which, weighing less than a ton, could destroy an urban area of more than ten square miles. In the war to which such an armaments race is likely to lead, the United States, with its agglomeration of population and industry in comparatively few metropolitan districts, will be at a disadvantage compared to nations whose populations and industry are scattered over large areas.

We believe that these considerations make the use of nuclear bombs for an early unannounced attack against Japan inadvisable. If the United States were to be the first to release this new means of indiscriminate destruction upon mankind, she would sacrifice public support throughout the world, precipitate the race for armaments and prejudice the possibility of reaching an international agreement on the future control of such weapons.

Much more favorable conditions for the eventual achievement of such an agreement could be created if nuclear bombs were first revealed to the world by a demonstration in an appropriately selected uninhabited area.

In case chances for the establishment of an effective international control of nuclear weapons should have to be considered slight at the present time, then not only the use of these weapons against Japan, but even their early demonstration, may be contrary to the interests of this country. A postponement of such a demonstration will have in this case the advantage of delaying the beginning of the nuclear armaments race as long as possible.

If the government should decide in favor of an early demonstration of nuclear weapons, it will then have the possibility of taking into account the public opinion of this country and of the other nations before deciding whether these weapons should be used against Japan. In this way, other nations may assume a share of responsibility for such a fateful decision.

Composed and signed by

J. Franck
D. Hughes
L. Szilard
T. Hogness
E. Rabinowitch
G. Seaborg
C. J. Nickson

Index

Index